TOO FAT?
TOO THIN?
Do You Have a Choice?

BY THE AUTHOR OF

Electric Fish

Sex Hormones:
Why Males and Females Are Different

Caroline Arnold

TOO FAT? TOO THIN?

Do You Have a Choice?

Foreword by Tony Greenberg, M.D.

WILLIAM MORROW AND COMPANY
NEW YORK   1984

*With thanks and appreciation to
Dr. Donald Novin and Dr. Carlos Grijalva,
Department of Psychology,
and Dr. Alfred Zerfas, School of Public Health,
at the University of California, Los Angeles,
for providing advice
and for reading and criticizing the manuscript.*

Text copyright © 1984 by Caroline Arnold.
All rights reserved. No part of this book may be reproduced or utilized in any form or by any means, electronic or mechanical, including photocopying, recording or by any information storage and retrieval system, without permission in writing from the Publisher. Inquiries should be addressed to William Morrow and Company, Inc., 105 Madison Avenue, New York, N.Y. 10016.

Printed in the United States of America.
1 2 3 4 5 6 7 8 9 10

Library of Congress Cataloging in Publication Data
Arnold, Caroline. Too fat? Too thin? Do you have a choice?
Bibliography: p. Includes index. Summary: Discusses the results of the latest studies and theories on weight control emphasizing the role of heredity as well as habits in determining individual weight. Includes a chart listing daily nutritional needs.
1. Reducing diets—Juvenile literature.
2. Body weight—Juvenile literature. 3. Fat—Juvenile literature.
[1. Weight control] I. Title.
RM222.2.A75   1984   613.2   83-23841
ISBN 0-688-02779-2 (pbk.)
ISBN 0-688-02780-6 (lib. bdg.)

Book design by Cindy Simon

Contents

Foreword, by Tony Greenberg, M.D.   ix

Introduction   1

1 **Everyone Needs Fat**   4
 WHAT IS FAT? / WHERE DOES FAT COME FROM? / THE IMPORTANCE OF FAT STORAGE / THE GROWTH OF FAT / WHEN FAT OR THE LACK OF IT CAN BE BAD

2 **Changing Your Body Weight**   23
 DIETS / EXERCISE / BEYOND DIETS AND EXERCISE: Drugs, Hormones, Diet Aids, Surgery / THINKING FAT OR THIN

3 **The Body's Resistance to Weight Control**   57
   IS FATNESS AND THINNESS INHERITED? / IS THERE A BODY-WEIGHT THERMOSTAT? / HUNGER

4 **Are You Too Fat? Too Thin? Or Just Right?**   74
   CHANGING ATTITUDES TOWARD BODY SIZE / MEASURING BODY FAT / TOO FAT / TOO THIN / DO YOU HAVE A CHOICE?

For Further Reading   93

Index   97

Tables

Calorie and Nutritional Content of
Common Foods   28

Daily Nutritional Needs   36

Calories Used During Exercise   41

Foreword

Obesity is known to be associated with a number of serious medical disorders, including heart disease, diabetes, high blood pressure, and orthopedic problems. Psychological effects often result in an unhealthy self-image and withdrawal from meaningful relationships. Social and job discrimination with economic impact are additional consequences. And in a culture dominated by concerns about physical attractiveness, it is no small wonder that Americans are so preoccupied with weight issues!

As writer of the "Dear Doctor" column in *'TEEN* Magazine for five years, I receive hundreds of letters each month helping me sense the pulse of teenage concerns. Letters regarding weight have always been most common. The majority are asking for

Foreward

practical advice on treating or avoiding obesity. Unfortunately, there is no quick answer for the teenager's need to be weight-balanced.

While anorexia nervosa and bulimia (gorging oneself followed by inducing vomiting) as life-threatening disorders requiring hospitalization are still rather uncommon, I am impressed that milder variants are increasing at an alarming rate. I certainly am receiving more letters asking about these extreme approaches to weight control.

It is clear to me that much of the problem in adolescent weight control stems from lack of basic knowledge about good nutrition and simple facts about obesity. Caroline Arnold fills this void. She has given us a handy reference to the principles of healthy nutrition, factors involved in appetite and eating behavior, and the underlying basis for obesity. Most importantly, the facts are presented in a practical way.

After facing so many questions on weight, it is a pleasure to have a book loaded with answers. The motivated reader will have many suggestions for achieving his or her weight goals.

TONY GREENBERG, M.D.
Associate Professor of Pediatrics
UCLA School of Medicine

Chief, Adolescent Clinic
Harbor/UCLA Medical Center

TOO FAT? TOO THIN?
Do You Have a Choice?

Introduction

What do you look like? Do you wish you were fatter or thinner? Do you wonder what you will look like when you finish growing?

During adolescence, that period when the body begins to change into its grown-up shape, you may find that you are becoming especially conscious of your appearance. At your age most people seem to want to look like fashion models or sports heroes, and those who don't often become dissatisfied with their bodies.

Most teenagers have normal, healthy bodies somewhere between the extremes of fat and thin. However, because they are surrounded by images of thin fashion models and muscular sports heroes in advertisements, movies, television, and in books and

Too Fat? Too Thin? Do You Have a Choice?

magazines, most think that they are too fat or too thin—whether they really are or not—and feel the need to change their weight. Do you? And when you try, do you find that it is not really easy and that your weight loss or gain is not always permanent? Do you wonder why? What you can do about it? That's what this book is about—the problem of weight control, why it is sometimes difficult, and what can realistically be done about altering body size.

When you want to reduce your weight, your first step is probably to cut back on the amount of food you eat. This may make you hungry, so you eat more food and then feel guilty. Thus, you become torn between two contradictory impulses—the need to eat and the desire to be thin. Being thin at the cost of hunger is not much fun, nor is it healthy. Everybody needs to eat and everybody needs some fat. Where fat comes from and how it is used in the body are explained in Chapter 1.

After you alter your diet, the second thing you may do is to change your amount of daily exercise. In addition, you may decide to seek help through such things as diet foods or diet pills. Chapter 2 discusses how diets, exercise, and other treatments can help you to gain or lose weight.

Nothing is more frustrating than going on a diet to gain or lose weight and not succeeding. Has this happened to you? Unfortunately, for almost all dieters, this is exactly what does happen. Although you may gain or lose a little at first, eventually your

INTRODUCTION

weight returns to its original level. Many experts now feel that body weight is controlled by a sort of weight thermostat which keeps it at about the same level all the time. Chapter 3 looks at some of the factors which may influence your body-weight setpoint.

The gigantically fat or reed-thin person clearly has too much or too little fat. But what about you? If you are at neither weight extreme, how can you tell if you are a little bit too fat or thin? Chapter 4 discusses the difficulties in measuring fat and in judging how much fat is appropriate for you. In addition, an annotated bibliography at the end of the book lists other books which provide further information on topics discussed in each chapter.

A healthy, attractive appearance does not have to conform to the rigid standard set by fashion or social trends. Your body shape and size is just as individual as you are. Like your personality they are part of what makes you unique.

This book will help you understand more about how your body works and why it may become fat or thin. It will also help you to understand how you can change your body shape, and to accept what cannot be changed.

1
Everyone Needs Fat

When Robert Earl Hughes of Monticello, Missouri, stepped on the scales in February 1958 he set a world record; he weighed an amazing 1069 pounds. Before he died several months later, he had reduced to 1041 pounds, but his body was still so large it had to be buried in a box big enough to hold a grand piano!

At the opposite end of the scale, there have been people of normal adult height weighing as little as 45 pounds. In their appearance they more closely resemble living skeletons than human beings.

Since most of us have weights far from these extremes, we may wonder how these people achieved their unusual body sizes. Did they choose to be fat or thin? Or were they in some way predestined to gain or lose enormous amounts of weight?

Questions like these have been asked for centuries.

People have attempted to control their weight chiefly through diets and exercise; but as any frustrated dieter knows, these methods are not always effective. Recent research has suggested that within the body a number of factors work to regulate body weight, keeping it within a certain range, so that it can be difficult, if not impossible, to alter your weight either above or below these limits.

To understand what these limits are and how your weight can be controlled within them, it is first necessary to understand the nature of body fat and how it is stored and used. Although your weight is determined by the sum of all your body parts, the chief variable is body fat. How much fat you have determines whether you will be fat, thin, or in-between.

WHAT IS FAT?

The sailors on the Kon Tiki are estimated to have weighed well over two hundred pounds each at the start of their journey—certainly not the model of beauty by today's standards! Yet had they been thinner it is possible that they would not have survived their long voyage across the Pacific Ocean. Their fat helped insulate them against the cold wind and water as well as providing a reserve source of energy.

Few people today embark on such arduous journeys as the members of the Kon Tiki group. Nevertheless, we all need fat in order to live. Among other things, fat helps to maintain body temperature, it

Too Fat? Too Thin? Do You Have a Choice?

cushions the bones and inner organs, and it is a compact store of reserve energy. It also aids in digestion by stimulating the flow of bile from the liver and emptying the gallbladder, where bile is stored.

The human body is composed of billions of tiny living cells, each performing a certain function. Fat cells, also called adipose tissue from the Latin word *adiposus,* which means "fatty," are designed to store and dispense tiny molecules of liquid fat called lipids. Lipids are formed during the process of digestion from the breakdown of the fats we eat. Other nutrients can also be converted to lipids within the body. The lipids or "simple fats" can be used by other body cells to produce heat and energy. The simple fats may also be used to construct more complex body fats such as cholesterol and lecithin, a process which takes place in the liver. Cholesterol and lecithin are important to the brain and nervous system. Cholesterol is also a source of bile acids and salts necessary for digestion, some hormones, and vitamin D.

Each fat cell is between 10 and 200 micrometers in diameter. Since one micrometer equals $\frac{1}{1000}$ of a millimeter, each fat cell is smaller than the size of a period on this page. There are from 20 to 160 *billion* of these cells in the adult human body.

Most fat cells are clumped together in a layer under the skin. They are yellowish-white in color and, if looked at under a microscope, appear to be porous and foamy. Like other cells, fat cells have an outer cell wall, and inside they contain cytoplasm and the cell nucleus. However, unlike other cells in the body, each fat cell is filled up with lipids. These lipids are absorbed from the bloodstream, and when needed by the body, they leave the fat cell and go back into the bloodstream to be carried to the liver or to the cells that need them. There is a constant exchange of lipids between fat cells and the bloodstream.

Fat cells can vary greatly in size. When one is empty of fat droplets, it collapses to almost nothing, just as a balloon shrinks when all of the air has been let out of it. However, like a balloon, each fat cell can expand to many times its original size to accommodate an increase in fat droplets. This ability of fat cells to expand and contract provides the body with a unique and flexible energy storage system.

Fat molecules are found in almost all tissues of the body. In addition to being found in fat cells, they compose 96 percent of bone marrow tissue and 22 percent of nerve tissue. Fat is found in the liver, blood, and in mother's milk.

Most of the fat we eat as well as the fat inside our bodies is a combination of three chemical forms of fat —olein, palmitin, and stearin. Each of these has a different melting point: olein melting at 0° C., stearin from 45° to 60° C., and palmitin from 55° to 70° C. The higher the proportion of olein in any fat, the lower its melting point. Each kind of animal has a body fat particular to its species. Human body fat, which contains from 67 percent to 80 percent olein, is liquid at 37° C. (the usual body temperature). Fats such as butter and lard, which are solid at room temperature, usually about 22° C., have less olein in them. When we eat fats such as butter or lard the body converts them to forms which can be stored in the body's fat cells. The digestibility of a fat depends on its melting point.

In humans and some other mammals, the amounts of body fat remain more or less constant after maturity. However, the amounts can vary from species to species, between sexes, and even among individuals. In some animals, such as those that migrate or hibernate, the amount of body fat changes considerably at certain times of the year.

Human beings are among the fattest of mammals. For most individuals, between 15 and 30 percent of their body's weight is fat, which is more than double the amount found in most other mammals. Whales are one of the few mammals with more fat than humans. Thirty-five percent of their body weight is devoted to fat.

Among people there are wide variations in the amounts of normal body fat. Women have proportionately more than men. They have an extra layer of fat under the skin which softens the angularity of the bones and gives the female body its soft rounded curves. Older people tend to have more fat than younger ones, and individuals who live in cold climates tend to be fatter than those living in places with warm temperatures.

Some animals, such as fish and camels, store their body fat in lumps. However, in most mammals, in-

Everyone Needs Fat

cluding human beings, fat is distributed rather evenly under the skin where it functions as a blanket between the inner organs and the outside world, helping to maintain a constant body temperature. At this temperature, usually 37° C., body cells function at their optimum level. The value of fat as insulation is obvious to any thin person who has shivered on a cold winter day while his or her fatter companions stayed warm. On the other hand, when the summer temperatures soar into the nineties, the thin person then has the advantage in keeping cool.

Even though all people store fat in the same places on the body, the amounts can vary a lot. Some people tend to store more fat on their upper body, and others on the lower body. Some may store it in the buttocks, others in the thighs. These areas of heavier fat deposit differ by age, sex, and race, and are largely hereditary.

Scientists have recently discovered a second kind of body fat which is brown in color. Brown fat, also called brown adipose tissue or BAT, comprises about 1 percent of all body fat in adults, although in babies it can be as high as 6 percent. Brown fat is both a better insulator and a more concentrated source of energy than white fat, providing up to twenty times the amount of energy as white fat cells. It is found in hibernating animals such as dormice and ground squirrels, in animals that live in the Arctic, and also in newborn mammals, including human beings, where it provides critical warmth during early life.

Too Fat? Too Thin? Do You Have a Choice?

Children and adults rely on body fat to keep warm, and when this fails they have a natural reflex which produces heat—shivering. Babies, however, do not shiver, but instead produce heat with their brown fat. It acts something like a self-fueled electric blanket under the skin. Although white fat cells must send their fat droplets to other cells to be "burned" for heat and energy, brown fat cells function as both storage and heat-manufacturing cells. This heat-producing quality of brown fat is aided by a nutrient called carnitine which is found in mother's milk.

By the time a human baby is one year old most of the brown fat has consumed itself and been replaced by white fat. The small amounts of brown fat which adults have are found between the shoulder blades, behind the breastbone, near the heart and kidneys, and in the neck. Because of its ability to "burn" fat to provide energy, scientists such as Dr. George Bray at UCLA think that the relative amount of brown fat in an individual may be a factor in obesity. Perhaps people with more brown fat are able to burn up excess fat more easily and thus avoid becoming overweight.

WHERE DOES FAT COME FROM?

Whether it is hamburger and fries, hot tamales, or sukiyaki, whatever we eat provides our bodies with the essential nutrients for life. If you consider the body to be a machine, food is the fuel on which it runs. However, the food we eat cannot be used directly by our bodies because the particles are too big.

Everyone Needs Fat

Also, in many cases the molecules are too complex to be used directly by the body cells. The process of chewing and digestion helps to break down these particles into pieces small enough to be transported by the bloodstream and to enter the cells.

The kinds of particles of which food is composed and which our bodies need are the body nutrients—carbohydrates (sugars and starches), fats, proteins, vitamins, and minerals. Our bodies use these nutrients for growth and repair, for the production of energy, and to aid the manufacturing of products within our cells that keep our body functioning.

The process by which food is converted into energy for the body is called metabolism. Comparing the body to an automobile, the primary "gas" on which the body runs is a form of sugar called glucose. Glucose either enters the body in foods such as fruits where it occurs naturally, or results from the breakdown of complex sugars, starches, fats, and proteins within the body. Normally glucose enters the bloodstream from the small intestine. Some of it may go directly to the cells that need it and some of it may be stored temporarily in the liver in a form called glycogen.

The liver acts as a kind of "energy thermostat" in the body. When the body needs more energy, the liver releases glucose into the blood; when too much sugar enters the blood, normally after a large meal, the liver converts some of it to glycogen and stores it. Then when the blood-sugar level drops (several hours after eating or during heavy exercise), the

Too Fat? Too Thin? Do You Have a Choice?

liver reconverts the stored glycogen to glucose and releases it. If the liver has no stored glycogen it can also form glucose from fats and proteins. The amount of glycogen that the liver can store is limited. If the body ingests more than this, the excess is stored as fat in fat cells until needed.

The body never wastes any potential source of energy. When you consume more food than your body needs, it sets it aside for later use. The extra fats you eat can be stored directly as fat, although during digestion they are broken into smaller pieces. However, extra proteins and carbohydrates cannot be stored in their original form, so the body converts them to fat. Once converted, they cannot be changed back again. Many dieters, in an effort to lose weight, avoid foods like fats and sweet desserts and eat more meats and vegetables. However, such a diet is useless if the dieter continues to consume more food than his or her body needs. All excess energy-containing foods, no matter what their source, will be converted to fat.

THE IMPORTANCE OF FAT STORAGE

The ability to accumulate fat cells and store energy was a critical step in the evolution of life. In the warm seas where the first single-celled plants and animals evolved, food was plentiful and surrounded the microscopic life-forms. Each living cell simply absorbed the necessary nutrients as it needed them for growth and energy. For these animals there was a constant exchange of food and waste. But as these simple life-forms became larger and more complex

Everyone Needs Fat

and began to move longer distances, it became inconvenient to have to eat continually. Thus animals and plants which could store energy for later use could move farther away from food supplies and so had the advantage in exploration and development.

The best way for an animal or plant to store energy is in the form of fat. Although small amounts of carbohydrates can be stored in the liver as glycogen, this organ cannot keep large amounts because, for every gram of glycogen stored, 3 to 4 grams of water must also be stored. On the other hand, fat can be stored with almost no water. A 180-pound man with 30 pounds of fat would weigh 300 pounds if the same amount of energy were stored as glycogen! In addition, the potential energy of fat is more than twice that of an equal amount of sugar.

In earlier times the ability to store fat also provided insurance against the unpredictability of nature. In cold climates such as that of northern Europe, food was plentiful during only a few months of the year. A good supply of body fat was essential in order to get through the lean winter months. Even in more moderate climates body fat helped protect against possible famine since people who could store fat more readily had a better chance for survival. However, today in those parts of the world where food is plentiful, the body's fat storage system has turned for some people from being a form of life insurance to being a liability.

The value of fat as stored energy can be seen in animals such as locusts, salamanders, and the many

Too Fat? Too Thin? Do You Have a Choice?

birds which migrate long distances. These animals build up enormous stores of fat before embarking on their journeys. One such long-distance traveler, a sandpiper called Wilson's phalarope, flies over five thousand miles each way on its annual journey between Mono Lake in the California Sierras to lakes high in the Andes of South America. After breeding in Canada, the birds gather at Mono Lake each summer to fatten up on brine shrimp. Many birds double their weight and some become so heavy that they can barely get off the ground when it is time to fly farther south. Once they leave, much of the trip is over open water where there is little food. Then the birds must rely on their stored body fat for energy.

Plants store fat in their seeds. This provides them with energy for germination and for early growth until a root system can be established. Many of our commercial oils are derived from seeds such as peanuts, corn, soybeans, and cotton which have a high fat content.

Animals that hibernate rely on stored fat to get them through the long winter. Black bears, for instance, spend the summer gorging themselves on berries, nuts, and other carbohydrate-rich foods in order to accumulate fat for the winter. By early fall when the bears begin to hibernate, most of them are downright plump. Yet by the time a bear comes out of hibernation in early spring it will have lost 15 to 30 percent of its body weight. For a 450-pound bear that means a loss of up to 135 pounds! A mother bear nursing young can lose even more. Studies of the ways bears utilize their fat during hibernation may result in medical advances for human beings. For instance, hibernating bears are known to have very high cholesterol levels. This condition is considered dangerous in humans because it may result in hardening of the arteries or in the formation of cholesterol gallstones. Yet bears do not seem to develop these problems. Learning how these conditions are prevented in bears may lead to new treatments for people suffering from them.

THE GROWTH OF FAT

Until recently a fat baby was considered to be a healthy baby and mothers filled their infants with formula and Pablum to achieve this result. Babies with big appetites were considered "good" babies.

The growth and multiplication of fat cells in the body is a normal process. However, one of the more important discoveries of the last two decades has been the idea that overweight babies and young children develop an excess of fat cells and therefore doom themselves to obesity for the rest of their lives. Studies begun in the 1960s by Drs. Jules Hirsch and Jerome Knittle of Mount Sinai Hospital and Rockefeller University in New York indicate that rats which had been overfed in infancy developed more fat cells than those which were not. Drs. Hirsch and Knittle suggested that, like the rats, overfed babies developed more fat cells in early life. This then predisposed them to become fat later on, because each fat cell, under normal conditions, stays moderately

Too Fat? Too Thin? Do You Have a Choice?

full of fat, so people with more fat cells will be able to store more fat than people with fewer fat cells even if each cell is filled to the same extent.

For a long time it was believed that fat cells only multiplied during childhood, but recent work by Dr. Lars Sjöström and a group of scientists from the University of Gothenburg in Sweden has challenged this idea. They have shown that fat cells divide and multiply whenever the body passes a certain maximum weight.

When a person consumes more food than necessary, the body converts it to fat. At first this excess fat will be stored in existing cells. Those cells will get bigger and bigger until they are many times their original size. If at this point the person goes on a diet and consumes fewer calories than the body needs, this stored fat will be used for energy and the fat cells will shrink to their original size. This is what usually happens during a period of temporary weight gain. However, someone who overeats for a long period of time will eventually fill all the fat cells to their limit. When this limit is reached the body will begin manufacturing new cells to accommodate the extra fat. These new fat cells are a permanent addition since weight loss does not reduce the number of fat cells in the body. A person who gains weight by adding new cells can never reduce to his or her original weight without shrinking all the fat cells, both new and old, to a level below normal. This can be done, but only at the cost of perpetual hunger. The body resists a state of reduced fat storage. When fat cells are underfilled, the appetite increases in an effort to refill them. Therefore, obesity either in early development or later should be avoided, for it carries a legacy which may not be removed.

This is not to say that children may not outgrow fatness. Although a clearly obese child is not going to become instantly slim as an adult, to some extent a child will lose the appearance of chubbiness as he or she matures. Certain periods of growth, such as babyhood and preadolescence, are characterized by fat deposition and would not be normal without it. A chubby baby usually lengthens out as a two-year-old. Most girls appear somewhat plump at the beginning of puberty. Yet by the time they reach their full height several years later, the plumpness usually disappears.

It has long been recognized that body weight is correlated with the onset of menstruation in girls. A girl usually has her first menstrual period when she weighs about 105 pounds, so larger girls often reach puberty earlier than smaller girls. However, recent studies done by Drs. Rose E. Frisch and Janet W. McArthur at Harvard University suggest that the critical component is not weight but the relative amount of fat in the body. Children have less fat in proportion to their total body weight than adults. The change to adult levels begins in adolescence and triggers puberty.

A girl needs to have 17 percent of her body weight as fat to have her first menstrual period and she

Too Fat? Too Thin? Do You Have a Choice?

must increase this to 22 percent to maintain regular ovulatory cycles. Body weight is also critical for sexual maturation in males and undernutrition can delay its onset.

Fat provides a reserve of energy needed for reproduction. During wars, famines, and other times of hardship, the birthrate drops dramatically. Such stressful circumstances may bring about weight loss, as well as a decrease in fertility in both males and females. Women who lose 10 to 15 percent of their body weight (about 30 percent of their fat) stop menstruating. No doubt this is a natural defense. If such an undernourished woman became pregnant her body would have trouble supporting the growing fetus.

Having too much fat can also interfere with a woman's menstrual cycles and her ability to become pregnant. All the parts of the body are interrelated so that when one of them is out of balance, the others are also affected.

Most women gain weight during pregnancy. Not only does the abdomen protrude but the whole body becomes larger and fatter. The average weight gain during pregnancy is usually about twenty-five pounds. Although approximately nine to eleven pounds of this is the fetus with its surrounding organs and fluids, and two or three pounds are accounted for by the enlarged uterus and breasts, the rest is body fat. This fat is accumulated in specific areas within the woman's body in preparation for the expected baby. During the nursing period following the baby's birth, a woman expends more energy than at any other time in her life. To stay healthy and to nourish her infant properly, she needs the energy that was stored as fat during pregnancy. Normally, by the time the nursing period is finished, most of this fat will be used up. However, some women seem unable to regain their girlish figures after childbirth and it is possible that the female hormones may contribute to this permanent weight gain. Women taking hormones in birth control pills often gain several pounds while on the Pill. The balance of hormones in the body is extremely complex and the influence of hormones on body fat is just beginning to be understood.

WHEN FAT OR THE LACK OF IT CAN BE BAD

Fat is essential for life, but having too much or too little of it can be both life threatening and unfashionable.

People who have too much fat carry more weight than their bodies were designed to handle. Thus, in doing even such simple kinds of muscular tasks as walking, lifting, or climbing, their muscles must work harder than if they weighed less, and this puts a strain on the heart. During muscular work heat is generated, but because fat is such a good insulator the obese person has difficulty losing this heat and this can result in discomfort and excessive sweating.

Because overweight people must work harder to exercise they become tired more easily. If they then avoid exercise, this lack of body activity can eventually interfere with other body functions such as digestion, breathing, and the circulation of the blood.

Obesity itself is not a disease, but a number of medical problems are directly linked to it. There is a greater tendency for obese children to suffer from respiratory illnesses and joint problems and to have more accidents than slimmer children. High blood pressure as well as higher levels of cholesterol and fatty molecules called triglycerides in the blood of fat children may lead to heart disease later in life. Most adults who are diabetics are obese. In fact, many of them were obese before becoming diabetic. Death from heart disease is much higher in obese adults than in slimmer individuals because the hearts of the obese are constantly overworked. Other disorders found more commonly in obese people are cirrhosis of the liver and gallbladder disease. In general, obese individuals are also more susceptible to the common cold, bronchitis, and pneumonia, as well as to postural defects and arthritis. It is no wonder that life insurance companies consider obese clients a bad risk.

In addition to health problems created by too much fat, the overweight can suffer from psychological and social problems as well. In general our society does not admire obesity. To many, fatness is ugly and evidence of slovenly habits and lack of self-control. Fat people are often teased or taunted, they find it difficult to buy clothing, they may be discriminated against when looking for a job, they may be given lower grades in school, and they may find it harder to make friends.

The dangers of being too thin are less widely recognized than those of being too fat. Short of absolute starvation, most people *want* to be thin in order to conform to the current Western fashion ideal of a tall, slender figure. In fact, moderate thinness has proved to be the healthiest weight to maintain. However, when thinness is maintained at the cost of constant hunger this can interfere with a person's physical and mental capacities. Even if you have never been on a weight-loss diet, you probably have experienced temporary hunger when there was no food readily available. In addition to craving something to eat you may have felt weak, nauseous, gotten a headache, and felt irritable. All of these are signals by the body that it objects to being starved. A person trying to stay thin on a reduced diet may experience all these feelings constantly. A lack of sufficient fat can also make a thin person more suceptible to illness by reducing his or her resistance.

Society often gives mixed signals regarding thinness. Being thin is preferable for wearing most fashionable clothes. However, when it is time to take off those clothes—at the beach or for a physical education class for example—a skinny body is admired less. Many thin people shun activities such as swimming because they, like the overweight, feel self-conscious with their bodies exposed to public scrutiny. For males being underweight can be partic-

Too Fat? Too Thin? Do You Have a Choice?

ularly difficult because thinness is often linked to weakness and can be considered unmasculine.

Although the amount of fat which is appropriate varies from individual to individual, good health is maintained when a person is neither too fat nor too thin.

2
Changing Your Body Weight

Jack Sprat would eat no fat,
His wife would eat no lean,
And so, between them both
They licked the platter clean.

Illustrations of this old nursery rhyme typically show a spindly Jack before a near empty plate seated across the table from his enormous wife who is hungrily contemplating her own heaping plate of food. The implication is obvious: people who eat too much get fat. Those who don't become thin. The word "obesity" is derived from the Latin words *ob* meaning "over" and *edere* meaning "to eat."

Yet food is only half of the energy equation that determines how much fat will be stored in the body.

Too Fat? Too Thin? Do You Have a Choice?

If there were a second verse to the rhyme about Jack Sprat and his wife, it might focus on Jack's passion for exercise as opposed to his wife's sedentary life. The use of energy for body maintenance and exercise is the other side of the energy equation and the amount of energy consumed or spent is measured in calories. In other words, to keep your weight level constant, the number of calories consumed must equal the number of calories spent. When this is not balanced there is weight gain or loss.

Unlike body height, which does not change substantially after maturity, body weight can change constantly. In fact, slight daily changes in body weight are normal. Most people, for instance, weigh a little less in the morning than at night. Seasonal changes may also be normal. Some people weigh more in the winter and less in the summer.

Although many factors affect body weight, daily eating and exercise habits are the most directly involved. If you routinely consume milk shakes, chips, cookies, or other calorie-rich foods after school or while watching television, this may well lead to your putting on a few extra pounds. On the other hand, a summer at sports camp with strenuous activities each day could result in some weight loss.

DIETS

How many times have you heard someone who has been offered food at a party or at a meal say, "No thanks, I'm on a diet," or "I really shouldn't . . ."? Millions of people go on diets every year. To most

Changing Your Body Weight

people the word "diet" means "weight-loss" diet. However, there are also individuals who diet to gain weight. In the United States alone over ten billion dollars a year are spent on weight control. Whether someone wants to gain or lose weight, an essential part of the diet's success is calorie counting.

Food is an organic substance, and like all organic substances, it is composed chiefly of four elements—carbon, hydrogen, oxygen, and nitrogen. When organic substances are burned the oxygen from the air combines with the carbon, hydrogen, and oxygen in the substance to form carbon dioxide and water, and energy is released in the form of heat and light. This process is called oxidation. Oxidation of organic molecules in the body is a complex chemical process, but, as in burning, the products are carbon dioxide, water (which we excrete), and energy. This energy produced by oxidation is measured in calories.

One Calorie is the amount of energy needed to raise one kilogram of water one degree centigrade, usually from 15° C. to 16° C. In order to find out how many Calories or how much potential energy a particular food contains, that food is placed inside a metal chamber surrounded by water. This mechanism is called a calorimeter. After being ignited by an electric spark, the food is burned and the heat produced is absorbed by the water. By measuring the change in the water temperature before and after burning and knowing how much food was in the chamber, the Calories per gram can easily be calculated.

Too Fat? Too Thin? Do You Have a Choice?

Using a calorimeter, scientists have found that carbohydrates, on the average, have 4.1 Calories per gram and that fats have 9.3 Calories per gram. Carbohydrates and fats are both completely usable by the body and there is nothing left over. However, when proteins are used by the body, several waste products (substances called urea, uric acid, and creatine) remain. Thus even though the average Calorie count for proteins in the calorimeter is 5.6 Calories per gram, in the body the same amount of protein can only produce 4.1 Calories.

Calorie charts, like the one on pages 28–33, show the Calorie count of common foods. If you want to find out how many Calories you ate in a day you can use this chart to find the number of Calories in each item, estimating if your portion is not equal to that listed. Serious Calorie counters use scales to measure their food exactly. Many food packages also list the Calories per serving of their contents. Then at the end of the day you can add the Calories from all the items together for your total. By keeping track of this for several days, you can find out how many Calories you eat per day on the average. Knowing this figure is useful if you want to increase or decrease how much you eat.

Calorie needs vary greatly from one person to another and depend on age, sex, height, weight, activity levels, and the climate in which the person lives. For instance, as shown on the chart on pages 36–37, a twelve-year-old boy who weighs 97 pounds and is 63 inches tall should eat about 2800 Calories a day, on the average. Yet a girl of the same age and weight

CHANGING YOUR BODY WEIGHT

who is only an inch shorter needs only an average of 2400 Calories a day. After the age of fifteen, when growth is nearly complete, a girl's Calorie needs are reduced to an average of 2100 Calories a day. For a boy, who is still growing, Calorie needs between the ages of fifteen and twenty-two can be 3000 Calories a day or more. In general, Calorie needs go down after the age of fifty. They go up with strenuous activity, and for women during pregnancy and lactation, the period of nursing after childbirth. Because each person's body is unique in its requirements, the best way to determine your own Calorie needs is to consult your doctor.

If you want to gain or lose weight you can decide where to add or subtract Calories in your diet by comparing the number of Calories in various foods on a Calorie chart. What you eat is just as important as how much you eat. For instance, whole milk has nearly twice the calories of skim milk. While they are virtually the same in their protein and carbohydrate values, skim milk has almost no fat. Fat makes the critical difference between a fried and a boiled egg as well, adding 33 Calories. Everybody needs to eat some fat for good health. The fats we eat are used in the body to construct other fats such as cholesterol and lecithin, and are necessary as solvents for vitamins A, D, E, and K. However, most people eat far more fats than they need. Foods that are high in fats are higher in Calories and easier than foods containing little fat for the body to convert into body fat. If you want to lose weight by eating fewer Calories, a good start would be to reduce fat in your diet. How-

Calorie and Nutritional Content of Common Foods

	Quantity
MILK AND MILK PRODUCTS	
cheese, cheddar	1 oz.
cheese, cottage (creamed)	1 cup
cream, light coffee	1 tbsp.
ice cream	1 cup
milk, skim	1 cup
milk, whole	1 cup
yogurt, plain	8 oz.
MEAT AND RELATED FOODS	
bacon (fried crisp)	2 slices
beans, red kidney (canned)	1 cup
beef, lean only (roast)	1.8 oz.
chicken, drumstick (fried)	2 oz.
egg, hard cooked	1
frankfurter (2 oz.)	1
hamburger, lean (broiled)	2.9 oz.
lamb chop (broiled with bone)	3.1 oz.
peanut butter	1 tbsp.
peanuts (roasted and salted)	1 cup
peas, split dry (cooked)	1 cup
perch, ocean (breaded and fried)	1 fillet
pork chop (broiled with bone)	2.7 oz.
salmon (canned)	3 oz.
sausage, pork link (cooked)	1 link
shrimp (canned)	3 oz.
steak, lean only (broiled)	2 oz.
tuna (canned in oil, drained)	3 oz.

Calories	Protein (grams)	Fat (grams)	Carbohydrate (grams)
115	7	9	trace
235	28	10	6
30	trace	3	1
270	5	14	32
85	8	trace	12
150	8	8	11
145	12	4	16
85	4	8	trace
230	15	1	42
125	14	7	0
90	12	4	trace
80	6	6	1
170	7	15	1
235	20	17	0
360	18	32	0
95	4	8	3
840	37	72	27
230	16	1	42
195	16	11	6
305	19	25	0
120	17	5	0
60	2	6	trace
100	21	1	1
115	18	4	0
170	24	7	0

Calorie and Nutritional Content of Common Foods (continued)

	Quantity
BREAD AND CEREALS	
biscuit, baking powder	1
bread, white (22 slice/loaf)	1 slice
bread, whole wheat (18 slice/loaf)	1 slice
cake, angel food	1 piece
cake, white; chocolate icing	1 piece
corn flakes (plain)	1 cup
macaroni (cooked)	1 cup
noodles, egg, cooked	1 cup
pancake, plain	1
pie, apple	1 slice
rice, white; long grain, cooked	1 cup
saltines	4
spaghetti, cooked	1 cup
waffle, 7″ diameter	1
VEGETABLES AND FRUITS	
apple	1
banana	1
beans, green (cooked)	1 cup
broccoli (cooked)	1 cup
carrots (cooked)	1 cup
corn	1 ear
cucumber	6 slices
fruit cocktail (canned)	1 cup
grapefruit	½
lettuce, iceberg	1 head
mushrooms (raw)	1 cup

Calories	Protein (grams)	Fat (grams)	Carbohydrate (grams)
90	2	3	15
55	2	1	10
60	3	1	12
135	3	trace	32
250	3	8	45
95	2	trace	21
155	5	1	32
200	7	2	37
60	2	2	9
345	3	15	51
225	4	trace	50
50	1	1	8
155	5	1	32
210	7	7	28
80	trace	1	20
100	1	trace	26
30	2	trace	7
40	5	trace	7
50	1	trace	11
70	2	1	16
5	trace	trace	1
195	1	trace	50
45	1	trace	12
70	5	1	16
20	2	trace	3

Calorie and Nutritional Content of Common Foods (continued)

	Quantity
orange	1
orange juice	1 cup
peach	1
pear	1
peas, green (cooked)	1 cup
potato (boiled and peeled)	1
potatoes, french fried	10 strips
strawberries (whole)	1 cup
tomato	1
watermelon	1 wedge
OTHER FOODS	
bouillon cube	1
butter or margarine	1 tbsp.
cola	12 fl. oz.
fat, vegetable (for cooking)	1 tbsp.
fudge, plain	1 oz.
gelatin dessert	1 cup
jams and preserves	1 tbsp.
mayonnaise	1 tbsp.
oil, corn	1 tbsp.
pickle, dill	1
potato chips	10
soup (beef noodle, canned)	1 cup
sugar, granulated	1 tbsp.
syrup, table	1 tbsp.

Calories	Protein (grams)	Fat (grams)	Carbohydrate (grams)
65	1	trace	16
120	2	trace	29
40	1	tracc	10
100	1	1	25
110	8	trace	19
105	3	trace	23
135	2	7	18
55	1	1	13
25	1	trace	6
110	2	1	27
5	1	trace	trace
100	trace	12	trace
145	0	0	37
110	0	13	0
115	1	3	21
140	4	0	34
55	trace	trace	14
100	trace	11	trace
120	0	14	0
5	trace	trace	1
115	1	8	10
65	4	3	7
45	0	0	12
60	0	0	15

Figures are from the U.S. Department of Agriculture.

ever, it would be unwise to eat extra fats to gain weight, because too many fats promote heart disease. Instead, it would be better to raise your Calorie count with a high protein and carbohydrate diet instead.

Different types of fats have different Calorie values. Animal fats tend to have more Calories than vegetable fats. They are also higher in saturated fats. Fat molecules, like other organic molecules, are made of long chains of atoms—chiefly carbon, oxygen, and hydrogen. When some of the links in the chain have multiple bonds, the fat is said to be unsaturated. But if the link is partially broken and hydrogen atoms are added (which happens to unsaturated fats at high temperatures) the molecule becomes saturated with hydrogen, or hydrogenated. Saturated or hydrogenated fats are more easily and more completely digested by the body. However, there is some evidence that eating more polyunsaturated fats reduces the levels of cholesterol in the blood, at least for some people. High levels of cholesterol in the blood are believed to cause hardening of the arteries, a precursor of heart disease. Therefore, much emphasis today is placed on eating polyunsaturated in preference to saturated fats. Good sources of unsaturated fats are corn, cottonseed, soya, and safflower oils.

Diets high in sugar also promote obesity although not to nearly the same level as high-fat diets. Studies have shown that rats which, in addition to the standard rat chow, were fed a selection of foods such as chocolate chip cookies, salami, cheese, bananas, marshmallows, milk chocolate, peanut butter, and sweetened condensed milk, gained weight. Those fed this so-called "supermarket" diet—a varied diet of easily obtainable calorie-rich foods—gained more than twice as much weight as rats eating only the nutritionally balanced but essentially boring diet of rat chow and water. Rats, like people, tend to eat more when good-tasting foods are available.

Sweet foods such as candy bars and soft drinks can be a source of quick energy because they are superloaded with sugar Calories, but eating too much of them can quickly lead to excess pounds. Foods such as pies, cakes, and candy are concentrated Calorie sources. A piece of apple pie contains 345 Calories and cake with frosting has 250 Calories. When compared to 80 Calories in a fresh apple or 65 Calories in an orange, the fresh fruit is the clear choice for a lower Calorie dessert.

In countries like the United States, the problem of obesity has significantly increased since 1900 and this has coincided with an enormous increase in the availability of processed foods containing sugar. Recent research suggests that a diet high in sugar increases the appetite for more sugar and thus contributes to obesity. Rabbits who were fed a sugar solution shortly before their regular feeding ate nearly twice as much as rabbits who were not prefed. If this effect is true for people as well, then a reduction of sugar in one's diet would be a significant aid in weight control.

It is important to follow the practices of good nu-

Daily Nutritional Needs for Maintaining Good Health*

	Age (years)	Average Weight (pounds)
CHILDREN	1–3	28
	4–6	44
	7–10	66
MALES	11–14	97
	15–18	134
	19–22	147
	23–50	154
	51 on	154
FEMALES	11–14	97
	15–18	119
	19–22	128
	23–50	128
	51 on	128

During Pregnancy

During Lactation

*Note: The Calorie and protein requirements on this chart are only a general guide to body needs. Since your weight, height, and energy needs may vary greatly from the average, you should always consult your doctor when trying to estimate your own body requirements.

Average Height (inches)	Calories	Protein (grams)
34	1,300	23
44	1,800	30
54	2,400	36
63	2,800	44
69	3,000	54
69	3,000	54
69	2,700	56
69	2,400	56
62	2,400	44
65	2,100	48
65	2,100	46
65	2,000	46
65	1,800	46
	plus 300	plus 30
	plus 500	plus 20

The figures in this chart are based on studies done by the U.S. Food and Nutrition Board.

Too Fat? Too Thin? Do You Have a Choice?

trition at all times, but particularly if you are on a weight-loss or weight-gain diet. In the search for a magic formula for weight control, millions of people look to new and often unsound diets which are constantly and readily supplied by authors in magazines and diet books. Some of these are crash diets promising that the reader will lose many pounds in a few days or weeks. Others severely restrict or overemphasize certain types of food. Diets like these can not only be dangerously unbalanced nutritionally, but misleading as well, because the encouraging initial weight loss is chiefly due to the loss of water in the body. As the glycogen in the liver is used up during the first stage of dieting, the water necessary to store the glycogen is used as well. As soon as the diet is stopped, the liver will again build up its store of glycogen and the water and pounds will return. Body fats, like other fats, do not dissolve in water, and contrary to the claims of some diet books, there are no foods which will "melt" or "dissolve" body fat.

A few popular diets which combine low-calorie intake with low carbohydrates have been severely criticized by doctors. Not only do the diets fail to provide many of the essential vitamins and minerals, but without sufficient carbohydrates the body cannot process the fats. The fats then produce chemical substances called "ketones" which act like a poison in the body. Low-calorie diets which rely on liquid protein supplements may also be harmful because many of these lack essential nutrients and in the extreme have even caused death.

Changing Your Body Weight

A recently publicized diet "revolution" is the high fiber diet in which the dieter eats a large amount of high fiber foods such as bran, nuts, fruits, and vegetables. Because the fibers of these foods are nonnutritive and are not digested, the stomach feels full without being full of calories. Although an increase in fiber content in most people's diets would do no harm in moderation, such a diet, like all diets, would only be effective as a weight-loss technique if the overall calories are below normal. It is hard to fool the body, and if fewer calories are ingested it will still feel hungry, even if the intestinal tract is full of fiber.

Weight-reducing diets severely limit the amount of food consumed and may be especially harmful to children. Although a reduced calorie count will slow the growth of fat deposits, it also interferes with the growth of the body's organs, including the brain. Growing bodies need sufficient nutrients and will suffer if they are not available in the necessary quantities.

The most successful weight-loss diets are those with nutritionally balanced foods and a reduced calorie count. These may not give such dramatic results at first, but, by causing the body to use its supplies of stored fat, they will provide longer lasting weight loss as long as the dieter sticks to them. A permanent change in eating patterns may be the most important step toward a better weight for both the habitual overeater and the habitual undereater. However, no diet should be undertaken without a doctor's supervision.

Too Fat? Too Thin? Do You Have a Choice?

EXERCISE

Both the athletes competing on a playing field and the spectators on the sidelines or at home in front of the television are using energy. The human body needs energy to perform every function, even during sleep. However, when we do physical exercise or strenuous activities and use our muscles, more energy and, therefore, more calories are needed. The calories that are being "burned" by the body are those which were consumed earlier in the form of food. One of the most common recommendations given to people who are overweight (in addition to the suggestion that they should eat less) is that they should exercise more.

Heavy work or sports can use many times the number of calories needed for seated work or sleeping. Even walking at 4 mph rather than 2 mph can nearly double the number of calories used. Increasing or decreasing the amounts of exercise can directly affect the amounts of body fat if food intake is kept the same. The chart on pages 41–42 shows the number of calories used during various activities.

Exercise by itself does not cause big changes in body weight. One reason is that one pound of body fat equals approximately 3500 Calories, and spending those 3500 Calories in exercise can be an arduous task. For a 100-pound person, that means 12½ hours of basketball, about 14 hours of walking at 4 mph, or about 6 hours of running at 7 mph. Larger individuals use more Calories in exercise because it takes more energy to move their larger weight. A regular

Calories Used During Ten-Minute Periods of Exercise

Activity	Body Weight (pounds)			
	100	125	150	175
badminton or volleyball	34	43	52	65
baseball (except pitcher)	31	39	47	54
basketball	46	58	70	82
bowling (nonstop)	45	56	67	78
canoeing—4 mph	71	90	109	128
chopping wood	48	60	73	84
cycling—5.5 mph	34	42	50	58
cycling—13 mph	71	89	107	124
dancing (moderate)	28	35	42	48
dancing (vigorous)	39	48	57	66
football	56	69	83	96
golfing	26	33	40	48
horseback riding	45	56	67	78
mowing grass (manual)	31	38	45	52
mowing grass (power)	27	34	41	47
Ping-Pong	26	32	38	45
preparing a meal	25	32	39	46
racquetball or squash	60	75	90	104
running—5.5 mph	72	90	108	125
running—7 mph	95	118	141	164
running—12 mph (sprinting)	131	164	197	228
shoveling snow	52	65	78	89

Calories Used During Ten-Minute Periods of Exercise (continued)

Activity	Body Weight (pounds)			
	100	125	150	175
sitting (reading or watching TV)	8	10	12	14
sitting (talking or writing)	12	15	18	21
skiing (Alpine)	74	80	96	112
skiing (cross country)	79	98	117	138
skiing (water)	47	60	73	88
sleeping	8	10	12	14
standing	10	12	14	16
swimming (backstroke—20 yd./min.)	26	32	38	45
swimming (crawl—20 yd./min.)	32	40	48	56
tennis	45	56	67	80
typing (40 words/min.—electric)	15	19	23	27
walking downstairs	46	56	67	78
walking upstairs	117	146	175	202
walking—2 mph	23	29	35	40
walking—4 mph	42	52	62	72

The figures in this chart are adapted from *The Partnership Diet Program* by Kelly D. Brownell (New York: Rawson-Wade, 1980).

Changing Your Body Weight

exercise program can help an overweight person lose about one-third of a pound a week.

A good exercise program is one that is done on a regular, daily basis. It can involve any kind of exercise you like and which fits into your schedule. Some people like to participate in team sports such as basketball or soccer. You may prefer individual exercise such as jogging or calisthenics. Even walking to school or to the bus stop can be good exercise if it is done at a brisk pace. Continuous exercise such as swimming, jogging, bicycling, or weight lifting are generally preferable for physical fitness to stop-start sports such as baseball or tennis. Whatever your physical activity, in order to benefit the body it must be long and vigorous enough to raise the body temperature, increase heart rate, and bring about heavy breathing and perspiration. Then exercise will not only cause a reduction in body weight, but it will firm flabby muscles, strengthen the heart, improve circulation and breathing, improve posture, increase flexibility of the joints, promote general physical efficiency and reduce chronic fatigue. In other words, exercise, like nutrition, is an essential component of good health.

It is especially important to exercise when on a weight-loss diet. When a person loses weight some muscle tissue is lost along with the fat because the body tries to replenish its depleted store of glucose using proteins from the muscles. But if regular exercise is included with a weight-loss diet, this muscle loss is minimized.

Too Fat? Too Thin? Do You Have a Choice?

People who exercise regularly increase the size of their muscular tissue. You only have to look at pictures in body building magazines to see the extremes to which this can be carried. Sports training also contributes to the development of muscles. For instance, runners develop strong leg muscles and swimmers develop strong arm and leg muscles. Since muscular tissue is heavier than fat tissue, however, someone who has changed his or her fat to muscle may not experience much change in body weight. Yet even without weight loss, a person who has exercised to be "in shape" will appear more trim because his or her muscles will be tight and smooth. By the same token, those who fail to exercise will lose some of their muscle tissue, for unused muscles deteriorate and become weak.

Exercise also affects the body's metabolism, or the rate at which it uses energy. Whenever someone goes on a reducing diet and eats less, the body's metabolism drops. This happens within two to four days after the beginning of the diet. This drop in the metabolic rate can be seen as a natural defense by the body. When it "realizes" it is getting less fuel, it suddenly becomes more efficient in using what fuel it has in order to conserve the amount that is left. When on a reducing diet a person's metabolic rate can decrease by 15 to 30 percent, thus making it much more difficult to lose the unwanted weight. Most people on prolonged weight-loss diets find that by the end of the third month very little weight is lost because by then the body has adjusted to its new lower level of food intake. Exercise, however, increases the metabolic rate, forcing the body to use its stores of energy, and thus is the best way to counteract this effect of dieting.

Exercise affects a person's appetite as well. A common belief is that exercise causes an increase in appetite, and, therefore, as a technique to lose weight, it is counter productive since you simply eat back the calories you just spent during exercise. However, some studies have suggested that under certain circumstances exercise actually decreases the appetite temporarily. For instance, in one study it was found that school children who had recess before lunch ate less than they did when recess was scheduled after lunch. Most studies of the relationship between appetite and exercise indicate that there is a considerable delay between the ceasing of exercise and the return of an increased appetite. One such study of British military cadets showed that after a bout of heavy exercise, the students' appetites did not increase until two days later.

There has been some controversy over the relationship between exercise and obesity. Dr. Jean Mayer at Harvard University studied the differences in activity levels between overweight and normal-weight children, and he found that overweight children exercise less. In one study he compared two groups of teenage girls. Both groups ate the same amounts of food, but the overweight girls exercised considerably less. In another study groups of elementary school children were given one extra hour of exercise each day. During this period none of the children became overweight, but as soon as this hour of compulsory activity was eliminated from the school day, the average weight of children in the group increased.

Other researchers, though, have shown that obese children are just as active as the nonobese. One study indicated that obese boys actually use more calories than normal-weight boys when they exercise.

Although children may well be active no matter what their body size, overweight adults tend to be less active than those who are not. It may be that this lack of activity is a result of obesity rather than its cause, since people who find it more tiring to exercise will tend to do it less often.

Today's society has become increasingly sedentary. People usually work indoors, often seated at desks; they ride when they could walk; and for entertainment they sit or lie in front of the television set. Although more people now recognize the need for exercise, for many the exercise is irregular or insufficient. Whether one is fat or thin or in-between, daily exercise is always part of a good health program.

BEYOND DIETS AND EXERCISE

For many, diet and exercise alone are not enough for complete weight control. Some people may attempt to alter their body chemistry through drugs or hormone treatments and by so doing change the effects of diet and exercise. For others, special diet foods help maintain a particular kind of diet. A few people resort to surgery for weight control. For everyone, taking a look at what motivates you to eat will give a clear view of the factors which may be influencing you to eat too much or too little, and thus to gain or lose weight.

Drugs

Every overweight person dreams of a magic reducing pill that would make weight loss instant and painless. Similarly, there are some overly thin people who dream of a pill to gain weight. Like Alice in *Through the Looking Glass,* a sip or a bite would solve all their problems. However, real medications, like Alice's imaginary "drink me" and "eat me" foods, can have unforeseen consequences.

Drugs can be used to suppress the appetite. Amphetamines and another drug called fenfluramine are "anorectic" drugs which lower a person's desire to eat. Although amphetamines are abused as street drugs, they can be used safely under a doctor's supervision to maintain a reduced weight. Even so, these drugs can cause side effects such as insomnia, excitement, dizziness, headaches, hallucinations, or hypertension. Fenfluramine, which acts as a depressant in the body, may cause drowsiness and diarrhea.

In 1982 a new product appeared that seemed to be the perfect solution for people who wanted to keep their weight down without starving. The chief ingredients of this product, a protein derived from kidney beans, interferes with the chemicals in the body which digest starches, thus preventing the body from making use of those calories. By taking one of these "starch blocker" pills at mealtime, the dieter could eliminate as many as four hundred calories per meal. However, many doctors doubt that these starch blocker pills would be as effective in the stomach as they are in the test tube, since the stomach contains other chemicals which could break down the starch blockers before they ever had a chance to block any starch. Also, there is the possibility that large amounts of undigested starch could create unpleasant side effects such as gas pains and diarrhea. Therefore, these pills will no longer be available until further research has revealed exactly how they work in the body and whether they are safe.

Hormones

The human body has over thirty-five different hormones, tiny chemical substances which direct almost every function of the body including growth, reproduction, and digestion. These hormones are produced by glands, and in the past glandular disorders or an imbalance of body hormones were often blamed for weight problems. Years ago thyroid hormone preparations were often used to help people lose weight in the belief that obese people had a lower than normal metabolic rate caused by an inactive thyroid gland. Conversely, it was thought that thin people had overactive thyroid glands. However, this has not proved to be the case. Experts now realize that hormones are only one of many factors affecting the growth of body fat and thyroid treatment is no longer popular.

Some hormonal imbalances, however, may result in weight problems. Diabetics suffer from an insulin imbalance and adults with diabetes are often obese. Insulin is secreted by the pancreas gland and is essential for the metabolism of glucose. Glucose cannot pass directly from the bloodstream into the cells that need it because the wall surrounding each cell provides a barrier between the inside and outside of the cell. Insulin makes it possible for glucose to break through this barrier. Insulin also helps fat molecules to enter fat cells. As an individual becomes fatter and fatter, his or her pancreas must manufacture an increasing amount of insulin to enable the excess fat to be stored. Because it is overworked, the pancreas eventually breaks down, and the individual becomes diabetic. Weight loss can cure this kind of diabetes in adults.

Another hormone that is associated with weight is adrenalin, also called epinephrine. Have you ever had an exciting or frightening experience just before a meal and then found that you had lost your appetite? This would have been the result of adrenalin. When blood-glucose levels are low, during emotional excitement or after an injury, the adrenal glands secrete adrenalin in order to raise blood-sugar levels. This provides instant energy and at the same time reduces hunger.

One little-known hormone with a long name,

Too Fat? Too Thin? Do You Have a Choice?

dehydroepiandrosterone, also known as DHEA, may prove to be beneficial in treating people who are overweight. This hormone seems to prevent the manufacture of fat cells in experimental animals prone to obesity. Treatment of obesity in people with hormones such as DHEA is a possibility for the future.

Some scientists feel that an imbalance of the sexual hormones may be the root of the severe weight loss associated with the disease anorexia nervosa. Ninety percent of all anorexia nervosa victims are female. Boys who become anorexic do so before puberty when their sexual hormones are at low levels. Girls with anorexia stop having their menstrual periods when they lose 30 percent of their body fat. Those who are successfully treated and who resume their monthly periods rarely have relapses. Since the sex hormones are the regulators of the different sexual functions in males and females, a disease such as anorexia nervosa which involves a breakdown of those functions and which affects the sexes differently may well be hormonally related.

Diet Aids

One of the dietary mainstays for people on low-calorie regimens is some type of artificial sweetener. Most people like to eat sweets, and today sugar is in breakfast cereal, candy, snacks, and desserts, as well as in many processed foods such as spaghetti sauce and peanut butter. Yet the extra calories from all of this sugar can be the downfall of anyone struggling to lose weight or to keep from gaining weight.

The discovery of low-calorie sweeteners such as saccharin and cyclamates have provided the means to have a sweet taste without overloading on calories. Saccharin, a chemical compound five hundred times as sweet as cane sugar, was discovered in 1879 by Constantin Fahlberg, a German chemist, and an American chemist, Ira Remsen. Cyclamates were another widely used sugar substitute, but since 1969 they have been banned because of possible harmful side effects. A new sweetener called aspartame is two hundred times sweeter than sugar. Unlike other artificial sweeteners which have a carbohydrate base, aspartame is derived from amino acids, chemicals that make up proteins. Thus it will especially benefit people such as diabetics who must severely restrict their consumption of sugar and other forms of carbohydrates. Although both saccharin and aspartame have been approved for human consumption, they carry health warnings and must be used with care.

Special diet foods may also be part of a weight-loss program. A tour through the aisles of any grocery store will reveal the proliferation of low-calorie, low-fat, low-sugar, low-sodium and other diet foods that are available. When these foods are used as a substitute for higher calorie foods they may help you to lose weight. Often these prepared foods are helpful because they tell you exactly how many calories each portion contains. With foods you prepare yourself you are often forced to guess the number of calories you eat, making it difficult to count calories.

However, even with special diet foods it is always important to remember that if you consume more calories than you need, you will still gain weight.

Surgery

In extreme cases of obesity, surgery can be used as a final resort. A 35-year-old woman from Portsmouth, Rhode Island, was one such desperate case. She was extremely overweight at 415 pounds and found it impossible to lose weight and keep it off simply by dieting. In order to limit the amount of food she could eat she had an operation in which surgeons closed off most of her stomach. Now she can only eat about one cup of food at a time before her stomach feels full. Such radical surgery will help her to reduce to a healthier weight by physically preventing her from overeating.

Another surgical procedure used on some severely overweight individuals, called an intestinal bypass, removes most of the stomach and intestine. The human body has sometimes been compared to a complicated elongated doughnut with the hole in the center being the gastrointestinal system. Food enters the hole at the mouth, passes through the stomach and intestine, and waste matter is excreted at the anus. All along the way nutrients are absorbed by the body. If part of the length of this tubular hole is reduced from twenty feet to two feet, nutrients have less opportunity to be absorbed. However, because an intestinal bypass is often followed by complications such as painful diarrhea, doctors are reluctant to perform this type of surgery except when absolutely necessary and then only for the most desperate cases of obesity.

A new surgical procedure called "suctioning" goes directly after the excess fat. Here doctors inject a salt solution directly into fat tissue. The fat cells break open and then the liquid fat is vacuumed out. This works only to a limited degree and there is a danger that in the process nerves can be damaged, leaving numb areas in the skin. Also, as a balloon which has been inflated for a long time looks wrinkled when the air is removed, in older people the skin may not shrink back to the new smaller size after suctioning.

Surgery is always the last solution to weight problems. It is far better to learn how to avoid the problems in the first place.

THINKING FAT OR THIN

Without food we could not survive, and usually we eat because we are hungry and our bodies need a fresh source of energy. But sometimes we eat for reasons that have nothing to do with hunger.

How many people habitually watch movies holding a box of popcorn? If the movie is compelling they may eat the whole box without ever really tasting one bite. When individuals eat without thinking, they may consume more calories than they need.

Other people may eat because they are unhappy. Since eating is normally a pleasant experience, they may overeat in order to provide themselves with a

Too Fat? Too Thin? Do You Have a Choice?

bit of pleasure. However, if they are unhappy because they are already fat, then the solution only adds to the problem.

By eating, someone may indicate allegiance to a person or idea, or show religious devotion. For instance, bread plays a part in many religious rites. For Christians, bread is broken at communion; for Jews, matzoh, a special kind of unleavened bread, is part of a religious celebration. The ancient Egyptians baked over fifty varieties and filled their tombs with loaves of bread so that the dead would have food in the afterlife. Today people all over the world bake and eat special kinds of bread at holiday times.

Some foods are consumed because they have become symbolically linked to the situations in which they are eaten. Wine drunk at a wedding celebration as a toast to the bride and groom becomes a gesture of goodwill; drinking coffee with a new neighbor can be a gesture of friendship; and sharing soft drinks at the local hangout can be a symbol of social acceptance for a teenager.

Different societies also dictate how much food should be eaten. In the United States it is customary to accept second helpings as a compliment to the host's cooking. It is also considered impolite to leave food on your plate, because this indicates that you did not like it. Thus, out of politeness many of us join the "clean-plate club" even after we are full. In some oriental societies, however, an individual is not considered polite unless he or she does leave some food on the plate—a comment on the host's generosity and bountiful resources.

Changing Your Body Weight

On the other hand, sometimes not eating, or fasting, despite feelings of hunger becomes more important than eating. Hunger strikes are occasionally a form of political protest. English women campaigning for women's rights at the beginning of this century used hunger strikes to attract attention to their cause. More recently Irish Nationalists have used the same technique and some, such as Bobby Sands, have even starved to death.

Many religions promote fasting for short periods as a sign of religious devotion and a means of purification. Sometimes a religion prohibits the eating of certain foods as well. For Roman Catholics the six-week period before Easter, called Lent, is a period of general sacrifice, and in some places such as the city of New Orleans, pre-Lenten festivals like the Mardi Gras provide the opportunity for a last minute indulgence before the fast. Lent begins on Ash Wednesday and the day before is appropriately called Fat Tuesday.

For whatever reasons food is eaten, most people develop their eating habits and attitudes toward food when they are young. These habits may be hard to break and can lead to weight problems later in life if they encourage eating too much or too little. It is easy to learn to overeat in today's world. Advertisements on television, radio, and in newspapers and magazines tempt and encourage young people to eat high-calorie food. Fattening foods are bought and kept in the house where they are too easily obtained. Junk foods are too commonly given as snacks and

Too Fat? Too Thin? Do You Have a Choice?

become preferred over nutritious ones. Too much time is spent on quiet activities such as television watching and not enough is spent exercising. Undue emphasis is placed on eating everything on the plate. Children, like adults, eat when they are upset or discouraged. Food is often used as a reward for good behavior, and trips to places like zoos and movies become occasions to overindulge in sweets and junk food.

Few people are immune from these pressures, but being aware of the things that influence eating behavior will help you to make choices about what to eat and when, and to avoid eating too much or too little. An essential part of any weight-control program is getting people to analyze their eating patterns and perhaps changing some of their attitudes toward food. Individuals eat for pleasure, to be with friends, to conform to society, to console themselves, to demonstrate wealth and power, to show love, and for countless other reasons. In any weight-control program it is important to first be able to recognize the difference between hunger and the social or psychological reasons for eating.

3

The Body's Resistance to Weight Control

Just as some people seem to be able to eat whatever they like and never gain an ounce, there are others who appear to gain weight just thinking about food. Although calorie charts are useful in calculating the number of calories in varying amounts of food, it is obvious that individuals differ in the amounts of calories their bodies need. Both inherited and environmental factors can affect people's appetites and whether they will become fat or thin.

IS FATNESS AND THINNESS INHERITED?

Long before the science of genetics was discovered, the common phrase "like father, like son" indicated

Too Fat? Too Thin? Do You Have a Choice?

that people recognized that parents pass on their characteristics to their children. Tall people tend to have tall children, curly-haired parents tend to have curly-haired children, fat parents tend to have fat children, and thin parents thin children. These and all physical characteristics are passed from one generation to the next by tiny structures in the cells called genes. There are thousands of genes in each cell, each controlling one or more aspect of physical development.

Just as there are genes which control height and eye color, there are genes which affect the body's metabolic rate and its ability to store fat. Animal breeders know that certain animals get fatter than others and that they do it in different ways. Pig farmers, for instance, choose Middle White pigs for the best hams, Large White pigs for the best bacon, and Berkshire hogs for fat. In cattle, poultry, and sheep, breeding is carefully controlled for relative amounts of fat as well as other characteristics.

The first proof that a "fat gene" was passed from one generation to the next was found in the 1920s in studies done on a strain of mice distinguished by a yellowish tinge to their coats and a strong tendency toward obesity. When full-grown these mice more closely resemble furry tennis balls than rodents. Three other strains of mice with fat genes have been discovered as well. A fat gene in rats was discovered in 1961 by two scientists, Lois M. and Theodore F. Zucker. These fat Zucker rats have become a convenient model of human obesity in studies of eating and weight gain.

The Body's Resistance to Weight Control

In human beings it has not been possible to isolate a single fat gene except in association with a few rare hereditary diseases which have other symptoms as well. The ability to store fat in humans is probably controlled by a number of genes interacting with each other.

Hereditary influences on body size can be seen in the propensity of certain groups of people to become fat or thin. Among some South Pacific island people, for instance, there is a distinct tendency toward fatness. For the Polynesians who sailed on the Kon Tiki, the ability to store fat was essential for survival. On the average these individuals weighed 250 pounds. Considering that the trip took three months and that they carried little food, the reserves of energy stored in their fat were critical.

At the other extreme are the Zulus of Africa. For them high fat storage would be disastrous since the survival of a Zulu who is living a traditional life depends on the ability to run swiftly in pursuit of wild game. Any extra baggage in the form of fat would slow him down. On the average, only 3 percent of a Zulu's body weight is made up of fat. This is far below the normal range of 15 to 30 percent. It seems that for the Polynesians and the Zulus their way of life and the environment favored certain body types. Those individuals whose body types conformed to the needs of their environment survived and passed on their genes.

Most people fall between the extremes of the Polynesians and the Zulus. Nevertheless, even among people whose ancestors came from less de-

Too Fat? Too Thin? Do You Have a Choice?

manding environments, there are variations in the ability to store fat. Today, in areas of the world where food is plentiful and people lead a more sedentary existence, a tendency to store excess fat may easily result in obesity. You have probably heard somebody say, "All I have to do is look at a piece of cake and I gain five pounds." Although this cannot be literally true, staying or getting thin is a challenge for people whose bodies are very efficient fat storing machines. For others, five extra pieces of cake wouldn't add an ounce to their body weight.

Just as people vary in their ability to store fat, they vary in the rate at which they burn calories for energy. Some people have low, supereconomical metabolic rates—they can get a lot of mileage out of very little fuel. Thus if they consume extra fuel it is stored as fat for later use. Others with high rates of metabolism are like gas guzzlers. High amounts of calories are burned in a short amount of time and usually little is left over for storage. Metabolic rate also varies with the size of an individual since it is related to heat loss. Since smaller individuals lose heat more quickly, they must have a higher metabolic rate to maintain their body temperature. Smaller animals, for instance, such as mice and birds, have higher metabolic rates than do people, and in general, smaller people have higher rates than larger people.

In an experiment done with volunteers at the Vermont State Prison, men were asked to deliberately overeat. Despite the fact that they each consumed approximately the same number of calories a day, their weight gain varied considerably—from 18 to 31 percent of their original body weight. Clearly some men gained weight much more easily than others. When it came time to lose those extra pounds, those who gained most easily found it hardest to lose.

On the average, one pound of fat is equivalent to 3500 calories. So, in theory, if you consume 3500 more Calories than your body needs, you will gain one pound. For instance, if over a period of 35 days you consistently ate 100 Calories too many (approximately one chocolate chip cookie a day) you would expect to become one pound heavier. Similarly, if you ate 100 Calories too little over a period of 35 days you could expect to lose one pound. This makes gaining and losing weight sound simple. Unfortunately it is impossible to calculate your daily Caloric needs that closely. They can vary greatly from day to day depending on your activity levels. Caloric needs also differ from one individual to another.

It is nearly impossible to separate the effects of the environment and genetics on body size. Our genes provide the blueprint for body growth but the environment guides the construction process. The fact that fat parents have fat children or that thin parents have thin children may be due to heredity, or that, as a family, they all have the same eating habits, or a combination of the two. Fat parents do not necessarily have fat children, nor do fat children necessarily have fat parents. Heredity and the envi-

Too Fat? Too Thin? Do You Have a Choice?

ronment interact to make each individual unique.

Whether or not your body reaches its genetic potential depends on good nutrition. If, for instance, you visit any of the American Colonial homes that have been turned into museums, you will be surprised by the shortness of beds and doorways. The average Colonist was much shorter than the average American today, and this was probably due to a less nutritious diet. From Colonial times to the mid-twentieth century, Americans have been growing bigger and bigger, although that trend seems to have stopped. With a nutritious diet available to nearly everyone, the full genetic potential of most individuals appears to have been reached. The same phenomenon has occurred in Japan since the end of World War II with the introduction of a higher-protein diet there. Contemporary Japanese children tend to grow much taller than their parents. Like height, body weight is also dependent on adequate nutrition.

IS THERE A BODY-WEIGHT THERMOSTAT?

Why don't people grow fatter and fatter? Scientists speculate that every individual has a certain set weight which his or her body attempts to maintain. If the body goes above this set weight the person's appetite will decrease or the energy output will increase until the set-point is reached again. The reverse will happen if body weight gets too low—the individual will begin eating more, or the body's metabolic rate will slow down to conserve energy.

This idea has been called the set-point theory and has been debated since the turn of the century when it was first proposed. A set-point is that point around which something is regulated. For instance, if a room thermostat is set at 72° F., whenever the temperature drops below that point, the furnace will turn on to bring up the temperature. If the temperature goes above 72°, the furnace will turn off. The set point of the furnace thermostat is 72°. The fact that most people tend to stay at one weight, give or take a few pounds, suggests that there is something in the body that monitors weight and deposits of body fat in much the same way that the thermostat for a furnace monitors room temperature.

Studies of rats show that, given free access to food, they will maintain a steady weight. If they are starved for a short period or artificially overfed, they will automatically increase or decrease the amount of food they eat during the period following the starvation or overfeeding and quickly return to normal. Genetically fat Zucker rats will maintain their weight in the same way that normal rats do. This suggests that, like the Zucker rats, some fat people's body weight thermostat may simply be set too high, causing them to maintain their fatness.

Several experiments with human beings bear out these findings. In one experiment volunteers purposely starved themselves until their body weights were reduced by about 25 percent. In another experiment volunteers purposely overate to increase their body weight by about 25 percent. In both groups it was much more difficult than had been expected for the individuals to gain or lose the weight. For some volunteers trying to gain weight, it was necessary to eat up to eight thousand Calories a day—more than three times a normal amount—to achieve even a modest weight gain. For volunteers in both experiments it was much easier to gain or lose weight in order to return to the original weight level.

In another study of severely obese people on a weight-loss diet a similar pattern emerged. All the individuals lost weight when Calories were severely reduced, but as soon as they resumed their normal Calorie intake their weight skyrocketed back to its original high. Like 95 percent of all weight-loss dieters, they lost the battle in the end and found that the low weight achieved through dieting was only a temporary state once they resumed their normal eating patterns.

In all these experiments the volunteers' original weights varied widely. Yet in every case that original weight was the standard to which the body returned after being artificially manipulated—either starved or overfed. Just as our bodies regulate temperature so that it goes neither much above nor below 98.6° F., it seems that there is also a setting for weight. However, unlike body temperature which is fairly uniform for all human beings, this weight setting varies widely from individual to individual.

Scientists have not yet located a specific weight-control center in the body, although it seems as if such a system must exist. It may be that there is no single center but rather that the body weight setpoint is determined by several systems in the body working together.

One place that scientists have looked for a weight control center is in the brain, particularly in an area called the hypothalamus. In addition to directing such automatic functions of the body as breathing and heart rate, the hypothalamus also influences eating behavior.

Part of the hypothalamus called the lateral hypothalamus (LH) has long been thought to be a feeding center in the brain. Experiments with rats show that if the LH is surgically damaged so that it can no longer function, those rats rapidly lose weight during the first three weeks following surgery when they reduce the amount of food they eat. The rats then maintain these new lower weights even when given free access to food. However, part of this may be due to the fact that animals with a damaged LH develop stomach ulcers. When these stomach ulcers are prevented the animals lose less weight and recover more quickly.

Another area of the hypothalamus called the ventromedial hypothalamus (VMH) affects weight in the opposite way. Animals which have had this part of the brain damaged begin to grossly overeat and soon become obese. It is as if this part of the brain which normally says "stop" to eating behavior can no longer get its message through. In rare instances, this overeating behavior has been found in people

Too Fat? Too Thin? Do You Have a Choice?

with tumors at the base of the brain where the VMH is located. Presumably the tumors prevented the VMH part of the hypothalamus from functioning.

The LH area of the brain seems to be an area which controls weight over short periods of time whereas the VMH area seems to be a long-term control center. Thus, the VMH area seems to play a larger role in regulating the body weight set-point and it does so by influencing the appetite. However, only when good tasting foods are available does a rat with a damaged VMH overeat. When given unpalatable foods, that rat will gain little or even lose weight.

Do chronically under- and overweight people then simply have their regulatory system maladjusted? If so, someday it may be possible to readjust it.

Studies of animals that hibernate suggest that the body weight set-point may be under constant adjustment. Woodchucks, for instance, become fat in the fall at the same time that their food intake actually decreases. In some way they become very efficient fat storers just before hibernation. However, when they wake up in the spring and resume their normal eating habits, their weight is maintained at a constant, nonfat level. Learning how the woodchuck adjusts its metabolic rate could provide clues to a means of adjusting that rate in people.

In both rats and human beings, daily food intake and amounts of exercise vary. Yet the body seems to make adjustments to allow for these differences. Around the turn of the century, the German scientist R.O. Neuman suggested that there is some internal mechanism for "burning" excess Calories. Using himself as an experimental subject, Neuman carefully counted his daily calories for a year. He found that his average daily amount was 1766 Calories and that by eating this amount his weight remained the same. During the second year of the experiment he added 433 Calories a day (for a total of 160,000 extra Calories per year), and in the third year he added another 204 Calories (for a total of 230,000 Calories more than his original "normal" amount).

Although he would have been expected to gain approximately forty pounds in the first year and another sixty pounds in the second year, amazingly, Neuman gained only a few pounds during the two years of overeating. How did his body get rid of the excess calories? Neuman suggested that when there is an increase in consumption the body simply "burns up" or wastes the excess Calories by producing more body heat. When calories are reduced, the opposite happens and body heat is conserved. He called this process "luxuskonsumption." The recent discovery of brown fat's ability to generate heat suggests one means by which Neuman's theoretical process could work.

HUNGER

The energy the body needs for heat and work is constantly being used up, so there must be a constant supply. This supply of energy is stored in the liver and in fat cells. However, like all storage systems, periodic replenishing is necessary when stocks fall too low. This occurs at each meal.

Why do we eat? Most basically we eat to stay alive. However, most people would say that they eat because they are hungry. The feeling of hunger is a signal from the body that it needs more food.

From the time of the ancient Greeks, people have tried to find out what determines hunger. One of the obvious symptoms of hunger is hunger pangs—moderately painful sensations caused by contractions of an empty stomach. Since 1940 scientists have recognized that these hunger pangs were related to the level of glucose (sugar) in the blood. When glucose levels fall and there is a reduced amount of body "fuel," hunger occurs. Yet the fact that diabetic patients have high amounts of glucose in their blood and still feel hungry tended to disprove this idea. However, in 1956 Dr. Jean Mayer of Harvard University suggested that hunger depended both on the amount of glucose in the blood and on the amount of insulin available to metabolize the sugar and make it usable by the body.

The next question concerned how the glucose and insulin affected the body. How did the body measure their levels? The first monitor of hunger appears to be the liver. Because this organ can sense the rate of metabolism and monitors the amounts of nutrients entering the bloodstream, it is the first to "know" that the body needs to eat. When you say "I feel hungry" you are responding to the lower amounts of glucose reaching your liver from the bloodstream.

The Body's Resistance to Weight Control

But do people eat only because they are hungry, or do they eat because there is food on the table? Do people not eat because their stomachs are full, or because they have "lost their appetites" for some other reason? These questions point to the basic dilemma faced by researchers of the behavior of eating. Both biological and psychological factors influence eating behavior and can indeed interact with each other.

The control of body weight would be simple if each individual only ate when he or she felt hungry and stopped eating as soon as he or she felt full. Being able to recognize signals from the body is the first step in knowing when and when not to eat.

However, many other factors influence how hungry we feel. Taste is a powerful influence on how much we eat. You have probably had the experience of trying a new food with a friend and finding that one of you liked it and the other one did not. Recently it has been discovered that the sense of taste for each individual is as unique as his or her fingerprints. Thus it is not surprising that food preferences vary from one individual to another. They also change as we grow. What we taste is determined chiefly by over 10,000 taste buds on our tongues which detect salty, sweet, bitter, and sour tastes. In small children the taste buds are closer together than in adults, and it is thought that children may taste foods more strongly than adults. It is also often true that tastes change as individuals grow.

Some of our taste preferences may reflect our an-

Too Fat? Too Thin? Do You Have a Choice?

cient heritage. Many anthropologists believe that the first human beings ate a mostly vegetarian diet much like that eaten today by chimpanzees and other primates. Their appetite for fruits and honey, relatively scarce items that were sources of energy, developed into a "sweet tooth," which we still have. With the wide availability of all kinds of sweets today, this craving for sugar can lead to both tooth decay and obesity.

You have probably noticed that when you have a cold your appetite diminishes. This is because your sense of smell is impaired. The sense of smell enhances taste and can therefore influence the appetite. We all know that the smell of a good food is enough to start the mouth watering. Other factors such as the appearance and texture of the food, whether we have previously tried it, and whether it offers a variety of tastes affect our appetite as well. It is harder to stop eating something that tastes good than something that tastes bad, even when we are not hungry. For instance, most people can always "make room" for dessert after a big meal. It is also easier to continue eating if there is a variety of foods to choose from and we can change tastes. Variety is the spice of life, but it may also be the road to obesity.

Social factors can affect hunger as well. In our society we eat three meals a day whether or not we feel hungry. Also, most people, and animals, like to eat with company. Even a chicken which is full after a large meal will begin to eat again if it sees another chicken eating nearby.

The Body's Resistance to Weight Control

Temperature is another factor affecting how much someone eats. Cold climates foster bigger appetites than warm ones, and in days before central heating, this extra energy was needed to stay warm. However, today excess calories may not be used in winter when people stay indoors and curtail exercise, and a gain in weight may result.

Your emotional state can also affect your appetite. A woman who was preparing to take a big examination commented that by the time her exam was over she would have gained or lost ten pounds. She knew that psychological stress usually affects people's eating patterns. Mild stress often causes an increase in appetite, whereas extreme stress usually depresses the appetite. Stress can be a big exam, a move to a new school or to a new city, an illness, a death in the family, or some other major life event which creates a disruption. In a high-speed society such as that of the United States today, daily life is filled with numerous pressures and stress. One of the easiest ways to alleviate this stress is by having something good to eat, because for most people, eating makes them feel good. Thus a bite of food can temporarily give a moment of pleasure during times of stress. At the same time it provides energy to cope with the problem. However, some people overdo the remedy and end up getting fat. This in itself may cause further stress and a vicious cycle can begin.

Are fat people hungrier than nonfat people? In certain cases, yes. Some overweight individuals

Too Fat? Too Thin? Do You Have a Choice?

seem to be hungrier as a result of having a higher body weight set-point. They need to eat more to maintain their greater number of body cells.

However, some fat people find it difficult to detect real feelings of hunger. In an experiment done by Dr. Albert J. Stunkard of the University of Pennsylvania on fat and nonfat people, the participants came to the laboratory before breakfast and were asked to swallow a small balloon which enabled the experimenters to monitor their stomach contractions. Then for the next few hours they were asked at various intervals whether or not they were hungry. The surprising result of the experiment was that, although nonfat people only said they were hungry when their stomachs were contracting, that is, when they felt hunger pangs, fat people said they were hungry even though their stomachs did not contract. In other words, they were not able to associate the feeling of hunger with stomach contractions. These overweight people seemed to be less "in tune" with their own body signals, and thus were more likely to overeat or to eat when their bodies did not need more food.

Another experiment, done by Dr. Stanley Schacter of Columbia University, indicates that fat people rely more on outside cues to determine hunger. When Schacter set the clock of the experimental room forward, overweight subjects felt hungry as soon as the clock registered mealtime, even though it was an hour or so early. Nonfat subjects did not get hungry until the correct time for a meal. More re-

The Body's Resistance to Weight Control

cent research has shown that fat people are not alone in relying on outside cues for eating. Certainly tasty food displayed on a television advertisement can make any viewer, fat or thin, feel hungry.

Feeling hungry and wanting to eat good foods are natural and normal responses to the body's need for energy. Yet in order to conform to the modern trend to look slim, many people put themselves into a perpetual state of hunger or, failing that, into a perpetual state of dissatisfaction with their body size. So many dieters "fail" to achieve their desired weight not through lack of self-control but simply because they are trying to push their weight beyond the limits set by their own bodies. It is important to realize that not all the factors which affect weight are under our control. The human body is an enormously complex machine with many interacting parts. As we continue to learn more about how this machine works we may discover which parts can be manipulated to change body weight and which parts are unchangeable.

4

Are You Too Fat? Too Thin? Or Just Right?

A visit to a crowded beach on a hot summer day will provide you with both a chance to cool off and the opportunity to observe the diversity of human body types. Some bodies seem destined to wear bikinis whereas others look more suited to potato sacks. Do some of these people need to change their body weight? And how can you tell if your own body weight is appropriate for you?

CHANGING ATTITUDES TOWARD BODY SIZE

Have you ever wished you were born in another place or another time? A moderately plump woman of today who struggles so painfully to acquire a pencil-thin figure may wish she had been alive one hun-

ARE YOU TOO FAT? TOO THIN? OR JUST RIGHT?

dred years ago when her figure would have been the height of fashion. Women with pear-shaped figures would have been perfect artist's models in fourteenth-century Flanders.

Every culture, past and present, has had its ideas about what the ideal man or woman should look like. Most have favored figure types considerably plumper than the present ideal. A famous Stone Age relic called the "Venus of Willendorf" depicts in stone the torso of an extremely rotund woman with enormous breasts and well-rounded hips. Even if this figure only represents the wishful fantasies of its carver, it nevertheless suggests that among the hunter–gatherer people of that time great value was placed on fatness, at least in women.

Even the more widely known Venus de Milo, depicting the Greek goddess of beauty, would never win a beauty contest today. If the statue were complete it would represent a woman slightly over five feet tall with body measurements of 37"-27"-38"! Clearly the ancient Greeks also preferred women of substantial size.

After the fall of the Roman Empire, Christianity rose in Europe and attitudes toward eating and body size changed. Gluttony was considered a sin and those who succumbed were threatened by the specter of hell as described so glowingly in Dante's *Inferno*. The roots of today's attitudes toward overeating as a sinful act are found in the Middle Ages. How many dieters, after indulging in a forbidden dessert or a second helping, say, "I feel so guilty"?

Too Fat? Too Thin? Do You Have a Choice?

The same critical attitude can be seen in today's slang term for overindulgence in food, to "pig out."

During the Renaissance in Europe large figures once again became fashionable. An examination of figurative paintings from the sixteenth century up to 1900 reveals the preference for large, fleshy women. From Rembrandt's *Bathsheba* to Renoir's *The Bathers,* plumpness in women seems to be a mark of beauty rather than a detriment. Even today a line of clothing for large women is called "More to Love," in recognition of the fact that at least from some points of view, bigger can be better.

In the past a large size was also acceptable for men. During the reign of England's King Henry VIII so many members of Parliament imitated the enormous size of the king that the chairs of Parliament had to be enlarged. In the nineteenth century Prince Edward of England changed fashion to fit his portly figure. When he found that he could only fasten one of the buttons of his suit coat, he declared this to be the new fashion. Even today men continue to button their suit coats only once even though several buttons may be attached.

After World War I dramatic changes in attitudes toward body size occurred. With the 1920s came the era of "flappers" and a trend toward boyish figures for women in both Europe and the United States. Although there was a brief return to more ample figures for women in the 1930s, as seen by movie stars such as Mae West, ever since the 1940s fashion magazines have featured ever taller and thinner models. One of the most well-known of these was a model of the 1960s nicknamed "Twiggy" because of her ultrathin body. Today both television and magazines continue to perpetuate the idea that slimmer is better.

If the hypothetical man from Mars obtained copies of current American fashion magazines before visiting Earth he would deduce that all American women were long-legged and slender, and that American men were large, muscular, and deeply tanned. Imagine the Martian's surprise after he landed and discovered that for every fashion-model figure type there were many more in all different shapes and sizes. If the Martian traveled around the world he would also find that the American view of a perfect figure was not shared by all. Among many African tribes, for instance, plumpness is considered not only to be a sign of beauty but also evidence of personal wealth and prestige, since only a prosperous family can afford so much food. Among some wealthy Indian families fatness is also a sign of elevated social position.

Society's standards for body size are an arbitrary measuring stick of beauty, and often rather narrowly defined. It is important to remember that good health does not always need to conform to high fashion.

MEASURING BODY FAT

For adults the most common way of determining whether or not someone is over- or underweight is to

Too Fat? Too Thin? Do You Have a Choice?

check that person's weight on a standard height and weight chart. Such charts indicate a range of acceptable weights for someone of a given height and age. People who weigh more than the highest weight in the acceptable range are probably overweight, and those below the lowest weight are most likely underweight. Unfortunately, this type of height-weight chart is not reliable for growing children.

The first standard height-weight charts were devised around the turn of the century by life insurance companies. Insurance companies had an important stake in this information, because they had learned from experience that overweight policy holders were more likely to die sooner and thus cost the insurance companies more money. To make the first charts the companies measured men who had bought life insurance between 1895 and 1900 and women who had bought life insurance between 1895 and 1908. (Few women bought life insurance in those days, so it took longer to gather the information they needed.) These charts were then used for nearly forty years.

Unfortunately these first charts contained several inaccuracies. For instance, the statistics came from those individuals who bought life insurance. But since this was a relatively small segment of the population in 1900 and composed mainly of wealthier people, it was not a good cross-section of the population. Also, no adjustments were made for variations in bone structure among individuals. Thus, for instance, a 5′5″ woman with small bones and a 5′5″

Are You Too Fat? Too Thin? Or Just Right?

woman with a large sturdy frame would be compared equally for weight.

In 1942–43 the Metropolitan Life Insurance Company revised the height-weight tables to make them more accurate. In the new tables the age scale was abandoned and people were divided into three groups—those with large, medium, and small frames. Unfortunately, no precise way of determining this was given so the individual was left to guess his or her own frame size. Another change in the new charts was the use of the less rigid word "desirable" for "ideal" weight.

However, even dividing people into different frame sizes did not solve all the problems, because it did not account for activity levels. For instance, in the case of men who work at heavy manual labor and for professional athletes such as football players, the extra weight of muscles would often push these individuals into the overweight category on the charts, thus making them ineligible for some kinds of life insurance. Although the person may have been overweight by comparison to others of the same height, he or she would not be overfat.

Many overweight Americans were cheered by the recent revision of the Metropolitan Life Insurance height-weight tables. On the average an individual can now weigh more and still expect to live a long life. Doctors attribute this change to the fact that many people now live healthier lives. They exercise more, many have given up smoking and attempted to lower their blood pressure, blood sugar, and cho-

Too Fat? Too Thin? Do You Have a Choice?

lesterol. The higher acceptable weights are no consolation for the fashion conscious, though, for while they are healthy, they are further than ever from the pencil-thin ideal of modern fashion designers.

Although height-weight charts have improved somewhat since they were first devised, and can be useful, other ways have been developed to try to get a more accurate measure of fatness.

One method of determining the amount of body fat in an individual is based on the fact that fat is less dense than nonfat body tissues. Less dense objects weigh less in water and float more easily than dense ones. Therefore, weighing an individual both underwater and out of the water, then comparing the weights, will indicate the relative amount of fat. However, this method is not always practical.

Another method to measure fat, and one of the easiest to do, is the skinfold test. Most human body fat is found directly under the skin. In places where it is loosely attached, such as at the back of the upper arm or at the waist, it can be easily measured with a device called a caliper. The caliper simply measures the width of a pinch of skin and then calculates the relative amount of fat. Generally, in adults a measurement of more than one inch of skin and fat in the arm suggests obesity. An adult can use this measure at home by pinching the skin of the upper arm with the fingers. Even without a caliper it is clear whether the amount pinched is much above or below an inch.

However, it is harder to determine obesity in chil-

Are You Too Fat? Too Thin? Or Just Right?

dren and teenagers than in mature adults by any of these means because of differing rates of growth and the different rates of fat deposit. In most mammals, including humans, body fat content increases during the body's development. Its rate of growth, however, is uneven. In the fetus, fat does not begin to develop until the fifth month of gestation. After this, fat cells continue to grow rapidly until the baby is almost two years old. Then the growth rate tapers off until puberty, when fat cells begin to grow rapidly again. Because puberty can begin as early as ten or as late as sixteen or seventeen, a wide variety in body size is typical among teenagers. In both boys and girls, weight continues to increase even after height starts to level off.

The wide variety in height and weight among the same age children can be seen on the graphs doctors use to record a child's growth. Such graphs compare your height and weight to that of other children the same age. Your doctor is concerned both with how much and the rate at which you grow. Both of these can help him or her to evaluate whether you are too fat, too thin, or just right. In general, taller people would be expected to weigh more than shorter people. But if, for instance, each time you are measured you are taller than average but not heavier than average, then you are probably thin. Conversely, if you are heavier than average, but only average height, then you could be overweight. Since so many factors affect whether you are over- or underweight, growth charts are only a first step in a doctor's as-

Too Fat? Too Thin? Do You Have a Choice?

sessment of your proper size. They are most useful when used over a period of time to follow your pattern of height and weight gain.

TOO FAT

Forty percent of the adult population of the United States is said to be overweight. Because of their bulk, they find it difficult to buy clothes, they may be discriminated against when applying for jobs, and may be socially shunned. Nobody likes to be fat. Yet finding the fine line that separates outright obesity from the heavy side of normal is difficult.

Some definitions of obesity allow only a 10 percent variation from the average normal or "ideal" weight, whereas others allow a 25 percent variation. Thus for a 130-pound individual obesity could begin at 143 pounds or 163 pounds.

A person must be careful not to mistake normal growth for the development of obesity. During periods of rapid growth, particularly in the preteenage and teenage years it is both natural and necessary to put on weight. However, if after full maturity is reached—in girls usually around seventeen or eighteen and in boys in the early twenties—pounds are steadily increased, then the individual's eating and exercise patterns should be carefully examined.

Statisticians have grouped fat and thin people into all sorts of categories. For instance, studies of large groups of children indicate that fatness, in part, relates to family income and race. In general children of poor parents tend to be thinner than children of wealthier parents. However, this pattern changes for teenage girls. Then girls from upper-income families tend to become thinner, perhaps in an effort to match themselves to fashion-figure models. In addition to differences according to income level, age, and sex, researchers have found, also, that on the average blacks tend to be fatter than whites, Catholics than Protestants, Baptists than Episcopalians, Eastern Europeans than Northern Europeans, Northerners than Southerners, and so on.

These kinds of generalizations do not predict whether an individual will be fat or thin; they only identify segments of society where obesity is more likely to occur. Although there may be certain situations which foster obesity, they do not cause it to occur in the individual.

TOO THIN

Becoming or staying thin is a compulsion of modern American culture. Yet defining thinness is just as difficult as defining obesity. Is thinness what your doctor says you ought to weigh, or is it the ability to wear a pair of size 6 designer jeans? According to charts such as those devised by life insurance companies, thinness would be weights below the range of normal for any given height. Just as a weight above the normal range increases an individual's chance of an early death, so do weights significantly below normal. However, there is some evidence that moderate slimness actually promotes longevity. Mice, for in-

Too Fat? Too Thin? Do You Have a Choice?

stance, that were underfed to keep their weight down actually lived twice as long as those that ate normally. In general, being thin is healthier provided it does not reach an extreme.

A recent headline in a women's magazine read, "You can never be too thin!" In some tragic cases people, usually women in their teens and twenties, who did believe such a thing literally died of starvation. Unlike political protesters or victims of famine, they suffered from a disorder called anorexia nervosa.

A girl whom we will call Janet was a typical anorexia nervosa victim. She came from an upper-middle-class family where she was a middle child. She had always been well behaved, responsible, and a good student in school. Believing herself to be too fat, even though her 135 pounds were not excessive for her 5'8" height, Janet began to diet. After unsuccessfully following numerous popular diets, Janet finally managed to lose 25 pounds. Despite the growing concern of her doctor and parents, Janet was deeply committed to her weight-loss program and progressively eliminated more foods from her diet until she was eating almost nothing. When other people pressured her to eat, she resisted fanatically. At her lowest point she weighed only 81 pounds and her life was in danger. Janet was lucky and found a therapist who was able to help her, and today she weighs a thin but healthy 115 pounds and leads a normal happy life.

Anorexia nervosa is not a new disease, but one

which has recently become more widely recognized. It was first described in medical literature in 1684 by an English doctor, Richard Morton. The patient was a seventeen-year-old girl whose menstrual periods had stopped, who had no appetite, whose body was so thin it appeared to be "a skeleton clad only with skin," and whose body felt cold to the touch. The patient, who continued to refuse to eat, died three months later of self-imposed starvation.

It is not easy to die of starvation. The body resists and will progressively consume itself in an effort to stay alive. The average person can survive without food, provided water is available, for about two months. It does this by breaking down its own cells for energy. The body depends on glucose for energy and after the limited supplies in the liver are used up, usually by the end of the first day of fasting, it begins to get this nutrient from proteins. Some of the first proteins to be used are the digestive enzymes in the stomach and intestine. Without these essential body chemicals food cannot be completely digested. It is important for victims of starvation to resume eating slowly to allow their bodies to build up these enzymes again. Otherwise they will suffer extreme gastric distress.

During starvation, after the enzymes of the stomach are used, the next source of protein is body muscle. However, the body cannot afford to use up muscle tissue exclusively because essential muscles in the heart, kidneys, spleen, and intestines would be quickly destroyed, so it turns to its stored fat for

Too Fat? Too Thin? Do You Have a Choice?

energy. Eventually the fat will be used up and the body will again turn to muscle tissue to get the needed glucose as long as it can. Many people who diet to get rid of their "ugly fat" don't realize that for every pound lost, only some of it is fat. The rest is muscle and other tissues in the body.

To become more energy efficient during periods of starvation, the body slows down some of its functions. The metabolic rate drops, the pulse slows down, blood pressure drops and the body temperature is lowered. Victims of both starvation and anorexia nervosa push the limits of their bodies' resources to the extreme in order to survive.

During World War II volunteers at the University of Minnesota purposely starved themselves so that researchers could find out how best to treat prisoners of war and other victims of starvation after the war was over. The volunteers ate a diet of only half their required calories while maintaining a normal schedule of exercise. They quickly lost weight and after the first few months had lost about half their body fat. By this time they had all become irritable and lethargic and had lost all interest in everything except their two daily meals. Under the stress of extreme starvation their bodies resisted all but the most necessary activities. The difference between the victims of war or the volunteers and an anorexia nervosa patient is that in the anorexia patient the starvation is self-imposed.

Sir William Withey Gull, an English doctor, gave

Are You Too Fat? Too Thin? Or Just Right?

anorexia nervosa its name in 1873. The word "anorexia" is derived from Greek and means "without eating." "Nervosa" indicates that the disorder seems to have its roots in the nervous system and is not a physical ailment.

Because the victim of anorexia nervosa chooses not to eat, the disease has traditionally been considered a psychological one, and treatment is usually by a psychiatrist. Some treat it as an individual problem and others as a family problem. Victims whose lives are in danger are hospitalized so their food intake can be supervised.

Nearly a quarter of all anorexic nervosa victims learn that one way to avoid getting fat is simply to vomit all the food they eat. Then they can eat freely or even gorge themselves with an overload of calorie-rich foods and then force themselves to regurgitate and therefore avoid the calories and subsequent weight gain. This habit, called bulimia or sometimes the "binge-purge" syndrome, is found in nonanorexic individuals as well. Like anorexia nervosa, it has recently received a lot of publicity and is therefore more widely recognized.

It is estimated that there are 100,000 anorexics in the United States alone. This is ten times the number of ten years ago. In part this increase is due to the disease being more widely recognized now and therefore identified more frequently by doctors and psychiatrists. Of those who are identified and treated for anorexia nervosa about two-thirds recover. Of those who remain, some die. Because anorexia ner-

Too Fat? Too Thin? Do You Have a Choice?

vosa has the highest death rate of any psychologically based disorder, doctors are eager to discover its roots in hope of preventing it and of finding a cure.

DO YOU HAVE A CHOICE?

For every individual the question of body size is personal. No single factor destines anyone to fatness or leanness. Genetics may dictate the level of the body's ability to store fat, a job or life-style may control the amount of energy that is expended in a given day, the climate may determine how much fat is needed for insulation, and a culture may dictate the kinds of foods that are available and the attitudes toward eating them.

In many respects you can choose your own body size, and in others you cannot. You can choose what and how much you eat. For instance, you can choose to do without an extra serving of dessert, to have a nutritious breakfast, or to eat a sandwich instead of a candy bar for lunch. By choosing low-calorie foods in preference to high ones or by choosing smaller portions of your favorite foods you can eat less. You may also prefer to eat artificial sweeteners or specially prepared diet foods to lower your calorie intake. If you want to eat more, you can do the opposite.

You can also choose where and when to eat—you can pass up the popcorn at the movies or limit your snacks at a party; and you can try to understand why you eat. People who eat without thinking usually eat more than those who restrict their eating to regular

Are You Too Fat? Too Thin? Or Just Right?

meals. If you learn to recognize the situations in which you eat, not because you are hungry, but because you are lonely or bored, or because it is the "thing to do," then you can choose whether or not to eat. Learning to recognize your body's hunger signals is useful for distinguishing between them and social pressures to eat.

You can also choose how much exercise you have each day and how to get this exercise—playing tennis or going bicycling, for example, instead of going to the movies. If you use more energy in exercise than you have taken in by eating then you will lose some of your body fat. If you use less, you will gain body fat.

Under a doctor's supervision, some people choose to use drugs or surgery in an effort to control their weight as well.

It is important to realize how much you can control your daily food intake and energy expenditure. Too many people become fatter or thinner than good health would permit out of negligence or apathy.

However, it is also essential to recognize that there are definite limits to which you can push your body size. To avoid disappointment and frustration you must have realistic goals for your body shape. You may just not be destined to be pencil-thin or pleasantly plump. Victims of anorexia nervosa develop a warped view of optimal thinness. Even when their bodies become dangerously thin, they feel that they are too fat and this drives them to even more drastic food reduction. With the help of your doctor you can

Too Fat? Too Thin? Do You Have a Choice?

find out what a reasonable body size for you should be.

Just as you cannot choose the genes that control your height, you cannot choose the genes that control whether you will tend to be fat or thin. Your basic body size is determined from the moment of conception, a result of the unique combination of genes contributed by your mother and father. Because half of your genes are the same as half of each of their genes, you are likely to be somewhat like either or both of them. Among the genes which they have passed on to you are those which affect growth and metabolism. These will determine your body's rate and pattern for storing fat. Your choice of foods and exercise can make small adjustments within this pattern, but a wide variation from your destined shape is not possible without major changes in other functions of the body.

One of the reasons that it is difficult to change your body weight is that there seems to be a mechanism inside each person's body which regulates how much fat can be deposited. This has been called the body weight set-point. Whenever your fat stores fall below or go above the set level, your metabolism changes to right the balance. Thus, no matter what your weight—fat, thin, or in-between—your body attempts to maintain it. The set-point theory helps explain why it is so difficult for most people to gain or lose weight. Although the set-point is probably determined genetically, it seems to be adjustable within certain limits. A permanent increase in your

ARE YOU TOO FAT? TOO THIN? OR JUST RIGHT?

amount of daily exercise can lower your body's setpoint by increasing the metabolic rate. Such exercise must be *in addition* to your regular exercise and must last for at least thirty minutes a day.

One of the best ways to keep fit, healthy, and to keep your weight at an appropriate level is to exercise regularly. For most people this takes a conscious effort. In a recent test of physical fitness among children ages six to seventeen, over half failed to perform at an acceptable level for good health. Many schools in recent years have cut back on their physical education programs and this has contributed to the drop in fitness among school children. Although community sport and exercise programs can make up for this, many people given the choice do not bother to utilize them, either because they are too busy with other activities or because they are not interested. If exercise is not already part of your daily schedule, then it is important to change your schedule to include some. Whether you choose to jog, dance, swim, play tennis or football, whenever you move your muscles enough to increase your heart rate and raise your body temperature you will be using calories that might otherwise have been stored as fat. And because exercise increases the body's metabolic rate, you will be able to eat more without gaining weight.

A good body weight is also maintained with a sensible diet. The most sensible diet for good health is one which is well-balanced nutritionally. If you include foods from each of the four basic food groups

Too Fat? Too Thin? Do You Have a Choice?

—vegetables and fruits; cereal products; milk products; and meat, poultry, and fish—you will get all the nutrients your body needs. It is especially important to eat sensibly during a weight loss or weight gain diet since a lack of or overdose of certain nutrients can be dangerous. In any diet, remember that there are no magical foods that will melt unwanted body fat or cause it to be used up more quickly. Neither "natural" or "organic" foods, nor special vitamins or food supplements provide any extra value in a weight control program, and it is best to be wary of any new "revolutionary" diet plans that make exaggerated claims about losing or gaining weight. Many of these are unfounded in fact and can be dangerous.

The best way that someone can keep his or her body weight at an appropriate and healthy level is by exercising daily and by monitoring the kinds and amounts of foods eaten.

Few people choose to be fat or thin. Yet an awareness of the many factors which affect body weight can help them to both accept the factors within the body which cannot be changed and to alter those things that can be changed.

For Further Reading

The following references were selected as representative sources for additional information on topics discussed in this book. Most should be available in your library.

ANTONACCI, ROBERT J. AND BARR, JENE. *Physical Fitness for Young Champions.* New York: McGraw-Hill Book Company, 1975.

This book defines physical fitness and helps you to evaluate your own level through a series of simple tests designed to measure individual skills. Specific advice is given for both boys and girls.

Too Fat? Too Thin? Do You Have a Choice?

BENNETT, WILLIAM AND GURIN, JOEL. "Do Diets Really Work?" in *Science 82*, March, 1982, pages 42–50.

This article discusses the body weight set-point and the scientific research which supports this theory. It also has a section which explains how height–weight charts were developed and the problems with using them.

BENZIGER, BARBARA. *Controlling Your Weight.* New York: Franklin Watts, 1973.

This book covers all aspects of weight control including diets and reducing exercises, and is aimed at both boys and girls who want to gain or lose weight.

BOSKIND-WHITE, MARLENE, AND WHITE, WILLIAM C. *Bulimarexia: The Binge/Purge Cycle.* New York: W.W. Norton and Company, 1983.

This book examines both the nature and treatment of the binge/purge cycle as well as the conditions in society which may encourage its occurrence. Although written for adults, this book has answers for anyone who wants to know more about this eating disorder.

EDELSTEIN, BARBARA. *The Woman Doctor's Diet for Teen-age Girls.* Englewood Cliffs, N.J.: Prentice-Hall, Inc., 1980.

For Further Reading

This book is a sound, practical guide to losing weight and keeping it off and is aimed specifically at girls ages 12–20. It includes helpful calorie charts and recipes. Although written for girls, most of the advice in this book would be useful to boys as well.

FODOR, R.V. *What to Eat and Why.* New York: William Morrow and Co., 1979.

This book focuses on the food nutrients, what they are, why we need them, and what happens when we do not get enough of a particular one. It also includes some information on food processing and suggestions for a healthy diet.

GILBERT, SARA. *Fat Free.* New York: Macmillan Publishing Co., Inc., 1975.

As its subtitle states, this book is a "common-sense guide for young weight worriers." It includes information about fat and sensible ways to keep from putting on too much of it. The book is written in a lively positive style and includes a good list of further reading at the end.

GILBERT, SARA. *You Are What You Eat.* New York: Macmillan Publishing Co., 1977.

This book, a "common-sense guide to the modern American diet," discusses what we eat and how our culture and the food processing industry affect our preferences. At the end is a helpful nutrition chart and a list of further reading.

Too Fat? Too Thin? Do You Have a Choice?

LANDAU, ELAINE. *Why Are They Starving Themselves?* New York: Julian Messner, 1983.

This book is specifically written for young people and it discusses the nature of anorexia nervosa and bulimia. It contains a bibliography for further reading.

LEVENKRON, STEVEN. *Treating and Overcoming Anorexia Nervosa.* New York: Charles Scribner's Sons, 1982.

Here, a noted authority on anorexia nervosa writes about the disease in a clear, concise, and nontechnical manner. The several case studies provide a dramatic focus on the problems involved in identifying and treating this disease.

Index

Africa, 77
adolescence, 17
adrenal glands, 49
adrenalin, 49
amphetamines, 47
anorexia nervosa, 50, 84, 86, 89
appetite, 45, 49, 57, 65, 66, 69, 71, 75, 85
artificial sweeteners, 51, 88
aspartame, 51

babies, 15, 17, 19, 81
binge-purge syndrome (*See bulimia*)
blood pressure, 79, 86

bloodstream, 7, 49, 68, 79
bodyweight set-point, 62–66, 72, 90
brain, 65
Bray, Dr. George, 10
bulimia, 87
calories, 16, 24–34, 36–42, 46, 48, 50, 51, 53, 57, 60, 61, 64, 67, 71, 86–88, 91

calorimeter, 25, 26
camels, 8
carbohydrates, 11–14, 26–34, 36, 51
children, 16, 17, 45–46, 78, 81, 91

Too Fat? Too Thin? Do You Have a Choice?

children, fat, 45–46, 58, 61
cholesterol, 6, 15, 20, 27, 34, 79
circulation, 43
climate, 71, 88
culture, 74, 75, 88
cyclamates, 51

Dante, 75
dehydroepiandrosterone (*See DHEA*)
DHEA, 50
diabetes, 20, 49, 51, 68
diarrhea, 47, 52
diet, 5, 12, 16, 21, 24, 25, 38, 39, 43–46, 52, 64, 84, 86, 88, 91, 92
diet foods, 46
digestion, 11, 48, 85
drugs, 46, 47, 89

eating, 53–56, 60, 65–73
energy, 5, 6, 11–13, 16, 18, 19, 23, 25, 40, 44, 45, 49, 53, 59, 60, 62, 67, 70, 71, 73, 85, 86, 88, 89
environment, 57, 59–61
epinephrine (*See adrenalin*)
exercise, 5, 24, 40, 43–46, 66, 71, 79, 82, 86, 89, 91, 92

Fahlberg, Constantin, 51
fashion, 74–77, 80
fat, 5–22, 26–33, 38–40, 44, 49, 50, 53, 58–63, 67, 74–77, 80, 81, 85, 86, 88, 89, 91
fat, brown, 9, 10, 67
fats, hydrogenated, 34
fats, unsaturated, 34
fenfluramine, 47
fertility, 18
fetus, 18
fish, 8
Flanders, 75
food, 21, 23–25, 27, 35, 38–40, 45, 50–56, 60, 65, 67, 73, 87, 88
Frisch, Dr. Rose E., 17

gall bladder, 6, 20
genes, 58, 59, 61, 90
genetics, 57, 61, 62, 88, 90
glands, 48
glucose, 11, 43, 49, 68, 85, 86
glycogen, 11, 13, 36
Greeks, 68, 75
growth, 11, 17, 27, 39, 48, 61, 69, 78, 81, 82
Gull, Sir William Withey, 86

heart, 85, 91
heart disease, 20, 28
heart rate, 43, 91

Index

height, 26, 36, 37, 78, 79, 80
height-weight charts, 78, 79, 80
Henry VIII, 76
heredity, 57–59
hibernation, 8, 9, 14, 15, 66
Hirsch, Dr. Jules, 15
hormones, 19, 46, 48, 49, 50
Hughes, R.E., 4
hunger, 16, 21, 39, 47, 49, 53, 55, 56, 67–73, 89
hypothalamus, 65

India, 77
insulation, 19, 88
insulin, 49, 68
intestinal bypass, 52
intestines, 52, 85

Jack Sprat, 23

ketones, 38
kidneys, 85
Knittle, Dr. Jerome, 15
Kon Tiki, 5, 59

lactation *(See nursing)*
lecithin, 6, 27
life insurance, 20, 78, 79
lipids, 6, 7
liver, 6, 20, 38, 67, 85

mammals, 81
Mayer, Dr. Jean, 45, 68

McArthur, Dr. Janet W., 17
menstruation, 17, 18, 50, 85
metabolism, 11, 44, 45, 48, 49, 58, 60, 63, 66, 68, 86, 90, 91
mice, 58, 60, 83
migration, 8, 14
Morton, Dr. Richard, 85
mothers, 14, 15, 18, 19
muscles, 43, 44, 79, 85, 86

Neuman, R.O., 67
nursing, 10, 18, 19, 27, 36, 37
nutrients, 10, 11, 52, 91, 92
nutrition, 28–33, 36, 37, 43, 62, 91

obesity, 10, 17, 19, 20, 23, 35, 40, 43, 45, 46, 50, 52, 53, 58, 60, 64, 65, 70–72, 77, 78, 80–83
overweight *(See obesity)*
oxidation, 25

pancreas gland, 49
plants, 14
Polynesians, 59
pregnancy, 18, 19, 27, 36, 37
Prince Edward, 76

Too Fat? Too Thin? Do You Have a Choice?

proteins, 11, 12, 26, 27,
   34, 36–38, 43, 47, 51,
   62, 85
puberty, 17, 50, 81

rabbits, 35
rats, 15, 34, 58, 63, 65, 66
rats, Zucker, 58, 63
Rembrandt, 76
Remsen, Ira, 51
Renaissance, 76
Renoir, 76
reproduction, 17–19, 48

saccharin, 51
Schacter, Dr. Stanley, 72
sex hormones, 50
Sjostrom, Lars, 16
skinfold test, 80
spleen, 85
starch, 11, 48
starvation, 21, 55, 63, 64,
   84–86
stomach, 52, 65, 72
Stone Age, 75
stress, 71
Stunkard, Dr. Albert J.,
   72
suctioning, 53

sugar, 11, 13, 34, 35, 50,
   51, 68, 70
surgery, 47, 52, 53, 65, 89

taste, 69, 70
temperature, body, 43,
   60, 67, 86, 91
thyroid gland, 48
tooth decay, 70
Twiggy, 77

underweight, 77, 78,
   81–88, 89

Venus de Milo, 75
Venus of Willendorf, 75
vitamins, 27, 38

weight, 16, 17, 26, 36, 37,
   43, 44, 51, 61–66,
   77–80
West, Mae, 76
woodchucks, 66

Zucker, Dr. Lois M., 58
Zucker, Dr. Theodore F.,
   58
Zulus, 59

613.25 Arnold, Caroline
ARN
　　　Too fat?　Too thin?
　　　Do you have a
　　　choice?

860212

DATE			
NO 26 '86			
MY 22 89			

TECHNICAL H.S. RESOURCE CENTER
ST. CLOUD, MN 56301

© THE BAKER & TAYLOR CO